Thomas Penz

Reconstruction of Magnetic Reconnection Events

Thomas Penz

Reconstruction of Magnetic Reconnection Events

Theoretical considerations and application to satellite data

Südwestdeutscher Verlag für Hochschulschriften

Impressum/Imprint (nur für Deutschland/ only for Germany)
Bibliografische Information der Deutschen Nationalbibliothek: Die Deutsche Nationalbibliothek verzeichnet diese Publikation in der Deutschen Nationalbibliografie; detaillierte bibliografische Daten sind im Internet über http://dnb.d-nb.de abrufbar.
Alle in diesem Buch genannten Marken und Produktnamen unterliegen warenzeichen-, markenoder patentrechtlichem Schutz bzw. sind Warenzeichen oder eingetragene Warenzeichen der jeweiligen Inhaber. Die Wiedergabe von Marken, Produktnamen, Gebrauchsnamen, Handelsnamen, Warenbezeichnungen u.s.w. in diesem Werk berechtigt auch ohne besondere Kennzeichnung nicht zu der Annahme, dass solche Namen im Sinne der Warenzeichen- und Markenschutzgesetzgebung als frei zu betrachten wären und daher von jedermann benutzt werden dürften.

Verlag: Südwestdeutscher Verlag für Hochschulschriften Aktiengesellschaft & Co. KG
Dudweiler Landstr. 99, 66123 Saarbrücken, Deutschland
Telefon +49 681 37 20 271-1, Telefax +49 681 37 20 271-0, Email: info@svh-verlag.de
Zugl.: Graz, Karl-Franzens-Universität, Diss., 2006

Herstellung in Deutschland:
Schaltungsdienst Lange o.H.G., Berlin
Books on Demand GmbH, Norderstedt
Reha GmbH, Saarbrücken
Amazon Distribution GmbH, Leipzig
ISBN: 978-3-8381-0410-2

Imprint (only for USA, GB)
Bibliographic information published by the Deutsche Nationalbibliothek: The Deutsche Nationalbibliothek lists this publication in the Deutsche Nationalbibliografie; detailed bibliographic data are available in the Internet at http://dnb.d-nb.de.
Any brand names and product names mentioned in this book are subject to trademark, brand or patent protection and are trademarks or registered trademarks of their respective holders. The use of brand names, product names, common names, trade names, product descriptions etc. even without a particular marking in this works is in no way to be construed to mean that such names may be regarded as unrestricted in respect of trademark and brand protection legislation and could thus be used by anyone.

Publisher:
Südwestdeutscher Verlag für Hochschulschriften Aktiengesellschaft & Co. KG
Dudweiler Landstr. 99, 66123 Saarbrücken, Germany
Phone +49 681 37 20 271-1, Fax +49 681 37 20 271-0, Email: info@svh-verlag.de

Copyright © 2009 by the author and Südwestdeutscher Verlag für Hochschulschriften
Aktiengesellschaft & Co. KG and licensors
All rights reserved. Saarbrücken 2009

Printed in the U.S.A.
Printed in the U.K. by (see last page)
ISBN: 978-3-8381-0410-2

Contents

1	**Introduction**	**3**
2	**Solar–terrestrial relations**	**7**
	2.1 The Sun and the origin of the solar wind	7
	2.2 Interaction of the solar wind with a magnetized planet	10
	2.3 The dayside magnetopause, reconnection signatures, and flux transfer events	12
	2.4 Polar cusps and continuous reconnection	16
	2.5 Structure of the magnetotail and implications for reconnection	18
	2.6 Global picture of reconnection in the Earth's magnetosphere .	23
3	**MHD equations and magnetic reconnection models**	**26**
	3.1 The basic equations	26
	3.1.1 The equation of continuity	26
	3.1.2 The equation of motion	27
	3.1.3 The adiabatic law	27
	3.1.4 The Maxwell equations and Ohm's law	27
	3.1.5 Simplification of the system of equations	28
	3.1.6 The reduced system of MHD	29
	3.2 Sweet–Parker model	30
	3.3 Petschek mechanism for magnetic reconnection	32
4	**Direct model of time–dependent Petschek–type reconnection in an incompressible plasma**	**36**
	4.1 Basic configuration	36
	4.2 Displacement vector in $\mathcal{L}$–$\mathcal{F}$ space	39
	4.3 Source term in $\mathcal{L}$–$\mathcal{F}$ space	41
	4.4 Displacement vector in real space	43
	4.5 Evaluation of the MHD quantities	46
	4.6 Magnetic field along trajectories	49
5	**Reconstruction methods and the inverse problem**	**52**
	5.1 Reconstruction methods for reconnection features	52
	5.2 The inverse problem in an incompressible plasma	57
	5.3 Reconstruction of an exponential reconnection electric field . .	60
	5.4 Reconstruction of sine–shaped electric fields	61
	5.5 Reconstruction of the reconnection site	64

6 Applications of the incompressible method to Cluster measurements in the Earth's magnetotail   **66**
   6.1   The Cluster mission 66
   6.2   Substorm and a series of NFTEs on September 8th 2002 67
   6.3   A NFTE on August 13th 2002 inside the plasma sheet 73

7 Extension to a compressible model   **76**
   7.1   MHD theory for a compressible plasma 76
   7.2   Application to the NFTEs on September 8th, 2002 82
   7.3   NFTE on September, 26th, 2005 84
   7.4   Analysis of high–latitude FTEs 87
       7.4.1   FTEs on 14th February 2001 88
       7.4.2   FTEs on 21th January 2001 92

8 Comparison between the analytical model and a numerical magnetotail simulation   **95**
   8.1   Simulation of non–stationary reconnection 95
       8.1.1   Simulator description 95
       8.1.2   Benchmark problems 97
   8.2   Simulation of reconnection in plain current sheet 99
   8.3   Reconstruction and comparison of the reconnection electric field 104

9 Green's function of compressible Petschek–type reconnection and associated wave phenomena   **109**
   9.1   Wave generation in the ambient plasma environment 109
   9.2   The case of symmetric magnetic fields 111
   9.3   The case of asymmetric magnetic fields 113
       9.3.1   Low plasma β case 113
       9.3.2   High plasma β case 115
   9.4   The appearance of side waves 119

10 Conclusions and Outlook   **122**

11 Appendix: The general Riemann problem   **125**

12 References   **131**

1 Introduction

Nearly all plasma flows, whether in laboratory fusion reactors, in our solar system or in the most distant regions of the Universe, generate magnetic fields (Hones, 1984; Priest and Forbes, 2000). The presence of these magnetic fields inevitably leads to the process of magnetic field line merging, which is triggered by a change in the conductivity of the plasma. This process is sub-divided into the so-called magnetic field annihilation and magnetic reconnection. The former is a slow dissipative energy conversion between the plasma and the magnetic field, appearing at a boundary separating two different plasma and field region. The latter is a topological re-structuring of the magnetic fields involved with drastic results: On one hand, stored magnetic energy is transformed into heat and kinetic energy of the plasma flow, which is a dominant source for free energy in a plasma. On the other hand, electric currents and fields are created, as well as shock structures, which affect the surrounding magnetic field.

Magnetic field line merging occurs or is thought to occur in a rich variety of plasma environments. In laboratory fusion machines such as tokamaks or in the Magnetic Reconnection Experiment (Yamada et al., 1997) it is evident that reconnection occurs. In the Earth's magnetosphere there are several phenomena which are attributed to be results of field line merging. But also at Mercury (Siscoe et al., 1975), the gas giants and in comet tails reconnection occurs (Priest and Forbes, 2000). Reconnection provides an elegant explanation for the motion of chromospheric ribbons and flare loops during solar flares (Priest, 1984). It is also proposed as a mechanism for the heating of solar and stellar coronae to extremly high temperatures of more than 106 K (Axford and McKenzie, 1996). Two astrophysical topics to which reconnection theory has been extensively applied are stellar flares and accretion disks, but also extragalactic jets, galactic magnetic fields, and even galactic clusters (Priest and Forbes, 2000).

The merging process was first applied to acceleration and heating processes in solar flares (Giovanelli, 1946). Dungey (1953) was the first to suggest that "lines of force can be broken and rejoined". But it took some more years, when Parker (1957) and Sweet (1958) were the first to introduce a simple MHD model for steadystate annihilation in a current sheet. Applying this model to solar flares shows that the rate at which magnetic energy is converted to kinetic energy is much too slow, therefore, the model is often referred to as a model for annihilation (or so-called slow reconnection). It was H. E. Petschek (1964), who developed a model, which predicts a reconnection rate close to the rate appearing in solar flares. Thus, Petscheks model was the first one of fast reconnection to be proposed. Today, it is believed to be

one of a family of more general models (Priest and Forbes, 1986).
Also time-dependent solutions of Petschek-type reconnection do exist. Semenov et al. (1983, 1984, 2004b) and Kiendl et al. (1997) have considered the temporal evolution that results, when reconnection is triggered by an increase in resistivity at a particular location in a current sheet (Erkaev et al., 2001). The onset of reconnection launches MHD waves into the plasma, according to the general Riemann problem (Heyn et al., 1988). In the incompressible version of the theory (Biernat et al., 1987, 1998), the fastmode waves propagate outwards instantaneously and set up an inflow of plasma towards the current sheet. During the socalled active phase of reconnection, a dissipative electric field is continuously generated in the diffusion region. The leading front of the outflow region is propagating along the current sheet at the local Alfvén velocity, therefore this region increases in size and very rapidly outgrows the size of the diffusion region. Since the outflow region plays the role of connecting magnetic field lines across the current sheet, thus establishing a topologically new region of reconnected flux, it may also be referred to as a field reversal region (Vasyliunas, 1975). At some stage, the reconnection process must switch off and the shocks detach from the site at which reconnection was initiated and propagate towards the edges of the current sheet. This model has also been extended to asymmetric conditions, which occur for example at the Earths magnetopause (Semenov et al., 1992). Several features and processes observed by various satellite missions are thought to be a clear evidence for magnetic reconnection in the Earth's magnetosphere. Satellite crossings of the diffusion region are considered to be a direct evidence for reconnection. Mozer et al. (2003) analyzed approximately 1000 subsolar magnetopause crossings, where they found about 75 examples of apparent electron diffusion regions. Vaivads et al. (2004) reported four–spacecraft observations of a diffusion region encounter at the Earth's magnetopause that allows to reliably distinguish spatial from temporal features. Also for the magnetotail, crossings of diffusion regions are reported (Øieroset et al., 2001, 2002). Another indication are accelerated plasma flows generated by reconnection. Paschmann et al. (1979) were the first to report the in situ detection of such flows. Sonnerup et al. (1981) did an extensive study of plasma flows due to reconnection by using the Walén and other tests. However, single–spacecraft observations revealed only single jets at the magnetopause. Phan et al. (2000) reported in situ two-spacecraft observations of bi-directional jets at the magnetopause using multi-spacecraft measurements. Energetic particles can act as tracers of the magnetic field topology and thus give an indication where the reconnection site is located (Daly et al., 1981; Daly and Keppler, 1983).
Since reconnection leads to the formation of outflow regions, also pertur-

bations due to the propagation of these bounded and accelerated plasma bulges can be detected and used to achieve information about the reconnection process. Russell and Elphic (1978) found a bipolar variation of the magnetic field normal to the magnetopause and deflections in the tangential components, which were interpreted as signatures of isolated tubes of magnetic flux. In the same year, Haerendel et al. (1978) found indications for reconnection processes also at the high–latitude dayside magnetopause. In the following years, a collection of studies supported the interpretation that FTEs are burst of magnetic reconnection at the magnetopause appearing most frequently during intervals of southward directed IMF (e.g., Rijnbeek, 1984; Kawano and Russell, 1997; Biernat et al., 2002). However, newer observations with the Cluster satellites show that for a northward–orientated IMF, reconnection can appear tailward of the cusp (Twitty et al., 2004).

Magnetic flux reconnected at dayside is convected into the tail lobes, where reconnection is initiated in a region between 15 and 30 R_e in the magnetotail. It was proposed by Sergeev et al. (1987) that FTE–like structures appear also in the Earth's magnetotail. In later work, these events were referred to as nightside flux transfer events (NFTEs) (Sergeev et al., 1992). These are short–term events in the substorm–time plasma sheet, which can be described by impulsive variations of the reconnection rate in models of transient reconnection. They are characterized, similar to FTEs, as a bipolar variation of the B_z–component and a deflection of the x–component of the magnetic field. Due to the extensive data from several satellite missions, different reconnection–associated structures were classified in the magnetotail. They are referred to as individual bursts of BBFs, as transient plasma sheet expansions, as plasmoids or flux ropes, as travelling compression regions, as well as NFTEs (Ieda et al., 1998; Nagai et al., 2001; Slavin et al., 2003; Sergeev et al., 2005).

After the observation of FTE signatures, several attempts were made to reconstruct and analyze different features of the reconnection process involved in the generation of FTEs. Early attempts were made by Southwood (1985), Farrugia et al. (1987), and Walthour et al. (1993, 1994). More recently, an fruitful approach to this topic was used by several authors (Hau and Sonnerup, 1999; Hu and Sonnerup, 2001, 2003) who developed a method based on the Grad–Shafranov equation to reconstruct two-dimensional space plasma structures in magnetohydrostatic equilibrium. Sonnerup et al. (2004) used this method to reconstruct the shape of a plasmoid. Recently, different models based on multi–spacecraft measurements were developed to determine the position of the reconnection site with respect to the satellite. Wild et al. (2005) used data from Cluster and Geotail positioned at the high– and the low–latitude dayside magnetopause. They consider the motion of reconnected flux tubes away from a user–defined location and compare the flux tube motion with the observed FTE–like signatures. In another approach, Fuselier et al. (2005) determine the reconnection inflow velocity into the magnetosphere, and use Cluster and IMAGE data to estimate the approximate reconnection site. A promising approach to reconstruct the reconnection rate, is the one described in this work. It is based on a time–dependent model of Petschek–type reconnection, allowing to take into account the temporal evolution of the flux ropes causing the FTE signatures. The solution of the direct problem of Petschek–type reconnection by Heyn and Semenov (1996) and Semenov et al. (2004a) is given in form of convolution integrals of the reconnection electric field and an integration kernel. In Laplace space, convolution integrals have a favorable form to solve inverse problems by using Tikhonov regularization. Using this method, it is possible to reconstruct the reconnection rate out of satellite measurements of the magnetic field (Semenov et al., 2005a, 2005b; Penz et al., 2006a, 2006b).

This work is organized as follows. In Section 2, the basic principle of solar–terrestrial relations are described especially emphasizing the different regions of the terrestrial magnetosphere were reconnection takes place. Section 3 gives a brief introduction to MHD theory and makes the reader familiar with basic models of magnetic reconnection. Section 4 gives the mathematical formulation of the direct problem of Petschek–type magnetic reconnection in an incompressible plasma, while Section 5 describes the inverse model and the reconstruction methods used in this work. Section 6 deals with the application of the incompressible model to several NFTEs observed by Cluster in the magnetotail. In Section 7 the theory is expanded to a compressible plasma, a profound comparison with the incompressible model is done, and applications to different FTE/NFTE observations are performed. Section 8 describes a theoretical investigation of wave phenomena caused by Petschek–type magnetic reconnection. The work is finished by a discussion of the achieved results and a detailed overview over relevant literature.

2 Solar–terrestrial relations

The solar–terrestrial environment is an extended region in the near–Earth space, the so-called geospace, including the upper parts of the terrestrial atmosphere, the outer regions of the geomagnetic field, called the magnetosphere, and the various emissions from the Sun, which are propagating through the interplanetary space and influencing the geospace. Starting from about 100 km above the Earth surface, the geospace extends to distances measured in tens or hundreds of Earth radii. It is a region where interactions and boundaries play are crucial role: interactions between terrestrial and solar particles, between solar and terrestrial magnetic fields, between magnetic fields and charged particles, and boundaries in form of current sheets between solar and terrestrial matter, and between regions of different patterns of flows.

The magnetosphere is the region of the geospace, where the geomagnetic field has a dominant influence. The form and the structure of the magnetosphere are determined largely by emissions from the Sun, while the magnetosphere reacts rapidly to changes in the solar emissions. Since solar activity is changing continuously, also the magnetosphere changes from day to day, from hour to hour, and on even smaller time scales. Therefore, we have to deal with a highly dynamical system, which is driven mainly by the variations of the Sun.

2.1 The Sun and the origin of the solar wind

The Sun is a G2 V star with a mass $M \approx 2 \times 10^{30}$ kg, which is located in the Milky Way. The visible surface, the photosphere, has a radius of about 700.000 km and black body temperature of about 5.800 K. Above the surface, two regions which are transparent to light can be found: the chromosphere and the corona. The chromosphere is a thin transition region extending 2.000 km above the photosphere. Considerably hotter than the photosphere, the chromosphere is heated by hydromagnetic waves and compression waves originated by spicules and granules. The temperature of the chromosphere is about 10^4 K. The corona consists of a hot gas merging gradually into the interplanetary medium, and flowing outward from it is the solar wind. Current theories indicate that the corona is heated by the dissipation of mechanical energy stemming from the convection zone, or by dissipation of magnetic energy by field-line reconnection (Axford and McKenzie, 1996). The kinetic temperature of the solar corona is about 10^6 K.

All transient phenomena occurring in the solar atmosphere are connected with magnetic fields leading to a 22–year solar cycle. Today all observed phenomena due to subsurface solar magnetic fields are inferred from the laws of magnetohydrodynamics. Sunspots can be considered as magnetic flux tubes with a mean diameter in the order of 1000 km. In sunspots the magnetic field lines are bundled and magnetic fields reach values of 0.2 to 0.3 T. But on the Sun, there exist much more phenomena associated with magnetic fields: faculae, granules, pores, flares, prominences, and so on (Hanslmeier, 2002). The mean magnetic field intensity measurable at the solar surface is only approximately 10^5 nT. The small–scale features of magnetic activity on the solar surface are continuously changing with a degree of randomness as a result of complicated turbulent and ordered convective motions in the envelope of the Sun. The large–scale sunspot cycle, however, shows a well–defined behavior as a result of convection and generation of poloidal and toroidal magnetic fields within the differentially rotating Sun. This magnetic field is the source of the so–called interplanetary magnetic field (IMF), which is an important factor for solar–terrestrial relations.

The existence of a continuous outflow of magnetized plasma from the Sun was a discovery which revolutionized space physics. This outflow was deduced by Biermann (1951), based on observations of cometary tails, which often have a dual nature: the long, sharply defined section known as the plasma tail that is always directed away from the Sun, and the more diffuse section called the dust tail which is curved. Biermann made the observation that solar radiation pressure alone was insufficient to account for the length of the plasma tail. Therefore, he postulated the existence of a material flow outwards from the Sun. This flow was assumed to be continuous, opposed to the transient flow proposed by Chapman and Ferraro (1930). During the past 50 years much more has been learned about the nature of this flow, which we call the solar wind, as a result of direct observations with ground–based instruments at a variety of wavelengths, and also from in–situ data recorded by a multitude of satellites at numerous location within the interplanetary medium.

If we consider the solar corona, the outermost layer of the solar atmosphere, whose mean temperature is about 10^6 K, it is hardly surprising that an outward flow exist, since the combination of the resulting thermal pressure and the magnetic pressure characteristic for the corona easily exceeds the restoring force of the gravitational field. Therefore, the outer layers of the solar environment are thus in dynamic rather than hydrostatic equilibrium, which was first described by Parker (1958). He used a model of an expanding corona with a radial velocity v_r,

$$\frac{d}{dr}(r^2 n v_r) = 0, \tag{2.1}$$

$$m_p n \, v_r \frac{d v_r}{d r} = -\frac{d p}{d r} - \frac{G m_s m_p n}{r^2}, \qquad (2.2)$$

$$p = 2 n k T, \qquad (2.3)$$

where p is the total pressure of electrons and protons, G the gravitational constant, m_s the solar mass, m_p the proton mass, n the particle density for a quasi–neutral plasma consisting of protons and electrons, k the Boltzmann constant, T the temperature, and r the heliocentric distance. Equation (2.1) is the continuity relation, Equation (2.2) is the equation of motion for a spherical symmetric flow, and the last one is the thermal equation of state. An additional assumption needed is a temperature profile with $T = const.$ until a certain distance, and $T = 0$ thereafter. This was the first model to give analytical solutions for the solar wind velocity corresponding to the work of Biermann.

The solar corona is widely believed as the source region of the solar wind, in particular the coronal holes. Coronal holes emit less light at all wavelength than adjoining regions, but it is most marked in the X–ray spectrum, where they appear as black areas. These regions are associated with an abnormally low plasma density, where the magnetic field has a single polarity – all inward or all outward. This crates open magnetic field lines, going out into the interplanetary magnetic field. The basic acceleration mechanisms are not well understood until now, but helmet streamer and explosive events are considered as important features. It is assumed that the acceleration process takes place between 2 and 5 solar radii, and ceases in about 20 radii.

The bulk flow velocity of the solar wind varies between 250 and more than 1000 km/s, according to the different types of solar wind, which can be distinguished (Schwenn, 1991). The sonic speed

$$v_S = \sqrt{\gamma \frac{p}{\rho}}, \qquad (2.4)$$

with $\gamma = c_p/c_v$ as the adiabatic coefficient and ρ as the mass density of the solar wind, and the Alfvèn velocity

$$v_A = \frac{B}{\sqrt{4\pi\rho}}, \qquad (2.5)$$

where B is the magnetic field strength and μ_0 is the induction constant, are much smaller than the bulk flow velocity. Therefore, the solar wind is a supermagnetosonic flow with Mach numbers $M_S = v_{SW}/v_S$ and $M_A = v_{SW}/v_A$ of about 10.

The solar wind consists almost entirely of protons and electrons in nearly equal quantities, which justifies the assumption of quasi–neutrality, with a small admixture of heavier elements, like 4 % ^{4}He^{++} and small parts of ionized heavier ions. This composition reflects the composition in the Sun's outer region.

An important consequence of the fact that the wind is a highly ionized, quasi-neutral plasma is that ideal magnetohydrodynamics may be used to model its behavior. From ideal MHD it follows that in a magnetized plasma, the conductivity goes to infinity, which eliminates the diffusive terms in the MHD equations. As a result, plasma elements remain adhered to the same magnetic field lines at all times, a phenomenon known as flux–freezing (Kippenhahn and Möllenhoff, 1975).

The Sun has a complex intrinsic field, and a remnant of this field, referred to as the interplanetary magnetic field (IMF), propagates into the heliosphere as a result of flux–freezing. Although the solar wind flow is radial, the same is not true for the IMF pattern. The solar rotation will force it to take the form of an Archimedean spiral, also referred to as a Parker spiral, since one end of the field line is rooted to the rotating solar surface, while the other is propagating radially outwards in the solar wind flow. Together with the tilted solar magnetic dipole, this leads to the so–called ballerina skirt model developed by Alfvèn (1977). As a result, the IMF will be inclined as a function of heliocentric distance r

$$\psi = \tan^{-1}\frac{\omega\, r}{v_{SW}}, \qquad (2.6)$$

where ω is the angular velocity of the Sun and ψ describes the inclination with respect to the radial direction. Another implication of the ballerina skirt model is the fact that the IMF can be directed either away from or back to the Sun, depending on the different sectors of the IMF.

2.2 Interaction of the solar wind with a magnetized planet

The solar wind may be treated as an ideal fluid under certain circumstances, and in general this is sufficient to describe the large–scale features generated upon encountering obstacles in its path, which include planets and their satellites, comets, and asteroids. In the broadest sense, there are three classes of solar system bodies that the solar wind must flow past: those that do not retain extended atmospheres, those that do, and those that possess both an atmosphere and an internally generated magnetic field. Since we are interested in solar–terrestrial relations, we only consider the latter case.

Before continuing the discussion of solar wind flow past the different classes of solar system bodies, we describe a feature common to many cases. Since the solar wind is highly supermagnetosonic no waves are able to propagate upstream from the atmospheric boundary to inform the incoming material flow of the obstacle. Thus, a shock must form some way upstream of the obstacle in order to deflect the IMF and flow. This is accompanied by a deceleration of the wind to a sub–magnetosonic level on the downstream side. Satellite missions to most of the planets have confirmed the existence of upstream bow shocks. The shock structures observed are very thin: typically a few ion gyro–radii. Since the mean free path in the solar wind is extremely long, this means that bow shocks can be classified as collisionless shocks. Wave–particle interaction have been identified as the primary dissipation mechanism, and the magnetic and electric fields within the shock can alter the ion and electron distribution functions sufficiently to drive microinstabilities and turbulence. These can change the particles velocities and therefore take the place of collisions as the dissipation mechanism (Schwartz, 1985). The greater part of the heating is thought to be due to wave–ion interactions. The extent to which parameter are altered across the shock are given by the well–known Rankine–Hugoniot conditions, which express conservation of mass, momentum and energy across the shock

$$[\![\rho v_n]\!] = 0, \tag{2.7}$$

$$\left[\!\!\left[\rho \mathbf{v} v_n + \left(p + \frac{B^2}{8\pi}\right)\mathbf{n} - \frac{\mathbf{B}_t B_n}{4\pi}\right]\!\!\right] = 0, \tag{2.8}$$

$$\left[\!\!\left[\rho v_n \left(\frac{v_n^2}{2} + H\right) + v_n \frac{B_t^2}{4\pi} - B_n \frac{\mathbf{B}_t \cdot \mathbf{v}_t}{4\pi}\right]\!\!\right] = 0, \tag{2.9}$$

where H is the enthalpy and brackets indicate the difference of a quantity upstream and downstream. Additionally, the continuity of the magnetic field normal component and the tangential component of the electric field must be satisfied

$$[\![B_n]\!] = 0, \tag{2.10}$$

$$[\![\mathbf{E}_t]\!] = 0. \tag{2.11}$$

It should be noted that as a result of both the parabolic shape of the bow shock and the inclination of the IMF to Sun–Earth line, the nature of the bow shock differs on the dawn and dusk sides.

By crossing the bow shock, the solar wind enters a region referred to as magnetosheath. The solar wind is slowed down to about 250 km/s and the corresponding loss of directed kinetic energy is dissipated as thermal energy, increasing the plasma temperature to about 5×10^6 K. Thus, the magnetosheath plasma is slower than the solar wind, but 5–10 times hotter. The flow in the magnetosheath around the magnetosphere was originally studied by Spreiter et al. (1966) within a purely gasdynamic approach. Such an approach is often used to describe the bulk flow of the magnetosheath plasma, predicting that the magnetic field strength and the density increase simultaneously as the magnetospheric boundary is approached. The addition of a frozen–in magnetic field in the frame of ideal MHD will lead to a decrease in the plasma density and a corresponding increase of the magnetic field strength. This leads to the problem, that in ideal MHD, at a stagnation point either the field strength increases to infinity ("magnetic barrier region") or the plasma density decreases to zero ("plasma depletion layer"), neither of which is physical (Erkaev, 1988). Pudovkin and Semenov (1977) showed that the development of a stagnation line instead of a stagnation point can circumvent the singularity. Crooker et al. (1984) argued that also a displacement of from the classical stagnation point towards dawn avoids the singularity. Additionally, one can imagine the situation where the plasma has piled up until a point is reached where the characteristic length scales lead to a breakdown of ideal MHD. Under such conditions, it is possible for the magnetic field to diffuse through the plasma, leading to magnetic reconnection.

2.3 The dayside magnetopause, reconnection signatures, and flux transfer events

The magnetosphere of a planet results from the interaction of the solar wind with the internal planetary magnetic field. The boundary between the solar wind and the Earth's magnetosphere is called the magnetopause. It is a current sheet whose thickness varies largely as one passes from the subsolar point to the flanks. The shape of the magnetopause is determined by the pressure equilibrium between the interplanetary magnetic field pressure, the solar wind pressure, the planetary magnetic field pressure and the plasma pressure. Using this pressure balance, it is possible to determine the stand–off distance of the magnetopause as (e.g., Grießmeier et al., 2004)

$$R_s \sim \left[\frac{M^2}{(\rho v^2 + B_{IMF}^2)}\right]^{1/6}, \qquad (2.12)$$

where M denotes the planetary magnetic moment, and B_{IMF} is the interplanetary magnetic field. As one can see from Eq. 2.12, the subsolar point is not static, but is determined mainly by the solar wind conditions ρv^2. It is located somewhere between 8 and 14 Earth radii with a mean at about 11 Earth radii.

As already mentioned above, there exists the possibility that ideal MHD breaks down and allowing the IMF and the terrestrial magnetic field to diffuse into each other, giving rise to the so–called diffusion region. Away from the diffusion region the plasma may still be treated as ideal and concepts like the frozen–in magnetic field still work. In the diffusion region magnetic reconnection is initiated. This is a process where two oppositely directed magnetic field lines approach, collide, break, and reconnect, leading to a fundamental rearrangement of the topology of the involved magnetic fields. Before reconnection, all field lines were either terrestrial dipole field lines (known as closed field lines, since both ends are rooted in the solid Earth), or IMF lines. But because of reconnection, a new class of field lines with one end still rooted in the Earth, while the other is extending into the interplanetary space, must be introduced. Such field lines are called open field lines. Several types of observations provide convincing evidence that reconnection takes place at the dayside magnetopause:

- Satellite crossings of the diffusion region: Mozer et al. (2003) analyzed approximately 1000 subsolar magnetopause crossings, where they found about 75 examples of apparent electron diffusion regions using data from the Polar satellite. They suggested the crossing of electron diffusion regions because their widths are several electron skin depths and the electron flow U_e within them does not satisfy the equation (in SI units) $E + U_e \times B = 0$. Vaivads et al. (2004) reported four-spacecraft observations of a diffusion region encounter at the Earth's magnetopause that allows to reliably distinguish spatial from temporal features. They found that the diffusion region is stable on ion time and length scales in agreement with numerical simulations. The electric field normal to the current sheet is balanced by the Hall term in the generalized Ohm's law, thus establishing that Hall physics is dominating inside the diffusion region.

- Accelerated plasma flows: Petschek–type models of magnetic reconnection predict that accelerated plasma flows should be generated. Such measurements were first reported by Paschmann et al. (1979). Also Sonnerup et al. (1981) did an extensive study of accelerated plasma flows at the dayside magnetopause. They applied the so–called Walén test to confirm that the accelerated flows are due to magnetic reconnection. This test is based on the fact that as a result of Maxwellian stress at an 1–D steady rotational discontinuity, the tangential velocity of the plasma changes from one side to the other according to the relationship

$\Delta \mathbf{v} = \pm \Delta \mathbf{v_A}$, where $\mathbf{v_A}$ is the Alfvén velocity and the sign depends on whether the observation was made north or south of the reconnection point. Thus, evidence for reconnection comes from the observation of accelerated flows which obey this relationship.

However, single–spacecraft observations revealed only single jets at the magnetopause, while the existence of a counter–streaming jet was implicitly assumed, since no experimental confirmation was available. Phan et al. (2000) reported in situ two-spacecraft observations of bi-directional jets at the magnetopause, finding evidence for a stable and extended reconnection line on February 11th, 1998. Equator–S observed high–speed northward plasma flows in many magnetopause crossings. The Geotail spacecraft, located about 4 Earth radii southward and about 3 Earth radii tailward of Equator–S, also crossed the magnetopause multiple times. However, in contrast to the jets encountered by Equator–S, those detected by Geotail were mostly southward. Nine of the 11 high–speed flows measured by Equator–S are northward while 14 of 16 flows measured by Geotail are southward. Thus, for the majority of cases, the reconnection site was located south of Equator–S and north of Geotail.

- Energetic particle signatures: Energetic particles can act as tracers of the magnetic field topology and thus give an indication where the reconnection site is located (Daly et al., 1981; Daly and Keppler, 1983). In a reconnection topology, magnetospheric particles can escape outward into the magnetosheath along open field lines. There they display a streaming distribution, with the preferred direction of steaming indicating the location of the reconnection site relative to the spacecraft. Such streaming particles are often observed in the magnetosheath, and they were called Magnetosheath Ion Flows (MIFs) by Neff et al. (1987). In the diffusion region, charged particles are accelerated directly to high energies through the reconnection electric field. These energized particles stream more or less along the separatrices bounding the reconnected flux region (Axford, 1984). Since multi–spacecraft measurements are available, it is possible to unambiguously resolve parallel currents along the separatrices and show that they are correlated with high-frequency Langmuir/upper hybrid waves (Vaivads et al., 2004). These waves can be involved in the thermalization of electrons and can be used as a diagnostics tool of reconnection sites.

- Flux Transfer Events (FTEs): FTEs were identified initially as characteristic magnetic field features seen at magnetopause crossings by

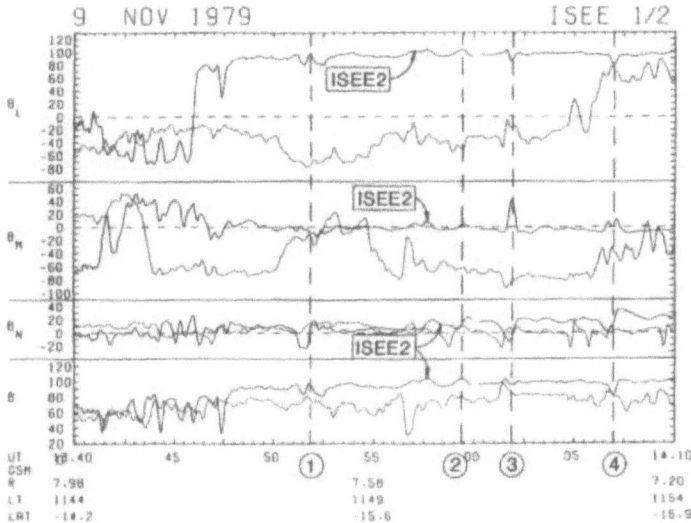

Figure 2.1: *FTEs (shown by dashed lines) seen simultaneously on either side of the magnetopause during a crossing by ISEE 1/2. FTE signatures are seen by ISEE 2 in the magnetosphere and by ISEE 1 in the magnetosheath (from Farrugia et al., 1987).*

ISEE 1 and 2 (Fig. 2.1). The magnetic signature consists of a bipolar variation of the magnetic field component normal to the magnetopause, simultaneous deflections of the field components tangential to the magnetopause, and a general enhancement of the total magnetic field strength. Russell and Elphic (1978) interpreted these observations in terms of a reconnected flux tube passing by a satellite (Fig. 2.2). In Fig. 2.2, one flux tube connecting magnetosheath field lines with magnetospheric ones is depicted, travelling northwards along the magnetopause. As it approaches the spacecraft, the magnetic field component normal to the magnetopause is displaced outwards, while as t recedes away from the spacecraft, it is displaced inward, thus giving the characteristic bipolar variation of the normal component. The occurrence of low–latitude FTEs is strongly correlated with the direction of the North–South component of the interplanetary magnetic field (IMF) with virtually no FTEs present when the field is purely northwards (Rijnbeek *et al.*, 1982; Southwood *et al.*, 1988). All these features imply

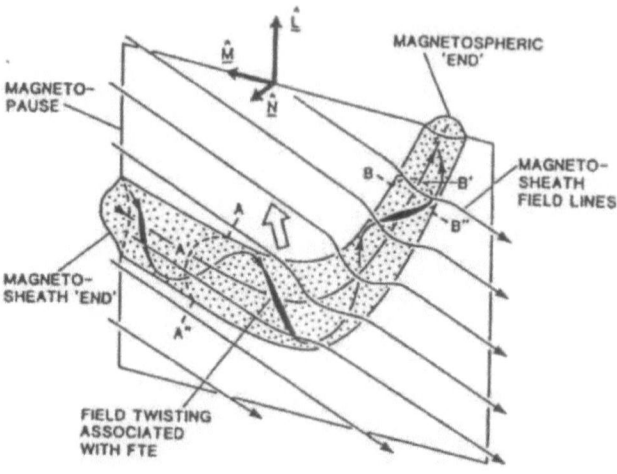

Figure 2.2: *Flux tube model of flux transfer events (from Russell and Elphic, 1978)*.

that FTEs are the result of time–varying and localized reconnection. The Petschek–type model of magnetic reconnection was generalized in a series of papers for a time–dependent reconnection rate (Biernat et al., 1987; 1998; Semenov et al., 1995). This model reproduces the bipolar B_N signature as well as the simultaneous deflections in the other field components, in good agreement with observations (Biernat et al., 2002). Kawano and Russell (1997) carefully studied the IMF dependence and source region of FTEs observed with the ISEE 1 spacecraft using an extensive data base comprising 1250 FTEs and confirmed that FTEs on the dayside magnetopause originate from reconnection. The suggestion that FTEs are a manifestation of time–varying and localized reconnection at the magnetopause has been verified by numerical, time–dependent MHD simulations of reconnection (Otto, 1991).

2.4 Polar cusps and continuous reconnection

Simple models of the magnetosphere predict two neutral points on the magnetopause where the total field is zero. These points connect along field lines to places on the Earth near ±78° magnetic latitude. These are the only points that connect the Earth's surface to the magnetopause and all the field from the magnetopause converges to those two points. They are therefore regions of great interest where solar wind particles from the magnetosheath can enter the magnetosphere without having to cross field lines. There is good evidence that this happens, since particles with energies typical of the magnetosheath are observed over about 5° of latitude around 78°, and over eight hours of local time around noon, forming so–called auroras. Being more extended than points, these regions are called the polar cusps.

Øieroset et al. (1997) presented observations of two types of auroral forms located at different latitudes in the cusp region. Type 1 (south) auroras are located at 71–75° magnetic latitude and occur during intervals of southward–directed interplanetary magnetic field. Higher latitude (77-78°) type 2 (north) auroras are associated with northward IMF. These observations are found to be consistent with the interpretation that type south auroras are a signature of low-latitude magnetopause reconnection, and that the type north auroras are associated with high latitude reconnection, tailward of the cusp. At high-latitudes reconnection occurs tailward of the cusp for northward IMF. Evidence of high–latitude reconnection consists of accelerated sunward ion flows in the lobes and of D-shaped distribution functions for the ions transmitted across the magnetopause, as established by in-situ observations (Gosling et al., 1991, Phan et al., 2003). Retinò et al. (2005) found similar evidence of magnetic reconnection on the duskside high–latitude magnetopause tailward of the cusp under mainly northward IMF. Important questions related to the large–scale nature of the reconnection are its continuity in time and the location of the X–line on the magnetopause. A key unanswered question, both theoretically and observationally, is how long the process can maintain itself naturally once initiated. In other words, is the process intrinsically intermittent or continuous? Intermittent, bursty reconnection turns on and off. Continuous reconnection operates at a variable rate but never ceases; if the fluctuation is a small fraction of the average then the reconnection is classed as quasi-steady.

FTEs and magnetospheric substorms - two phenomena believed to be initiated by reconnection - are highly burst–like occurrences, raising the possibility that the reconnection process is intermittent, storing and releasing magnetic energy in an explosive and uncontrolled manner. However, different observations indicate that reconnection can be also a continuous process. At low latitudes for southward IMF in–situ evidence of long lasting reconnection flows on a time scale of few hours has been interpreted in terms of quasi-steady reconnection (Gosling et al., 1982; Phan et al., 2000). Indirect evidence of continuous reconnection has been obtained from proton aurora measurements by Frey et al. (2003) with northward IMF and from radar

measurements by Pinnock et al. (2003) with southward IMF.
Another important issue is the location of the X–line, the place where reconnection is initiated. According to large–scale models of reconnection, the location of the X–line is controlled by the relative orientation of the IMF and the Earth's magnetic field. Different models predict different locations of the X–line. In the antiparallel merging model (e.g., Crooker, 1979) the reconnection occurs in localized regions at the magnetopause where the magnetic fields are nearly antiparallel. On the other hand, in the component merging model (e.g., Gonzalez and Mozer, 1974) the reconnection can occur even if only one component of the magnetosheath and magnetospheric magnetic field is oppositely directed. In general, it is difficult to conclude whether the component or the antiparallel merging model best describes the large-scale configuration of magnetic reconnection at the high–latitude magnetopause without knowing the location of the X–line, which has been inferred in a quantitative way using measurements in the Earth's magnetospheric cusp (Fuselier et al., 2000) and in a qualitative way using proton aurora measurements (Fuselier et al., 2002; Phan et al., 2003). On the basis of cusp aurora observations, Fuselier et al. (2002) concluded that under northward IMF high-latitude reconnection can be better explained by antiparallel merging. In this description the location of the X–line is limited to a localized region on the magnetopause and it depends on the value of the IMF B_y component i.e. the X–line moves to the dawn/dusk flank of the magnetopause for any finite dusk/dawn IMF B_y component. Nevertheless, recent observations show that at high latitude with northward IMF antiparallel and component reconnection can occur at the same time (Trattner et al., 2004). In this interpretation reconnection occurs at the high-latitude magnetopause also where it is not predicted by the antiparallel merging model, the only difference being that the process is less efficient than in the antiparallel situation. Retinò et al. (2005) reported two examples of X-line encounters detected using the observations of ion jet reversals. One of them has low magnetic shear which is inconsistent with antiparallel merging predictions.

2.5 Structure of the magnetotail and implications for reconnection

A magnetotail may be defined as a region in space created by the interaction between the intrinsic magnetic field of a large–scale object and a flowing magnetized plasma. It is in the downwind region of the object and contains a magnetic field configuration in which the field is oriented predominantly along the direction of the flow of the incident plasma. The Earth's magnetotail is form by the outflowing plasma from the Sun interacting with the Earth's intrinsic magnetic field. The interaction results in an electric current system that distorts the geomagnetic field and creates a field configuration in which the field points predominantly parallel or antiparallel to the incident solar wind flow. The Earth's magnetotail may be considered to be an extension of the atmosphere, elongated in the downwind direction, not only because the magnetic field in the magnetotail can often be traced back to the Earth, but also because a significant portion of the particle population in the magnetotail originates from the Earth.

A large portion of the magnetotail consists of two low–density region known as the tail lobes, one in the northern half of the magnetotail and the other in the southern half. The field lines threading through the tail lobes are also open. The energy density in the tail lobe is still dominated by the magnetic field. Bordering the tail lobe at its lower latitude interface is the plasma sheet boundary layer. This region is often the most dynamic plasma domain of the magnetotail, where ion beams coming from the Earth and from further downstream are often found. It is also where a lot of plasma wave activities in the magnetotail are detected. Magnetic field–aligned currents, flowing toward or away from Earth, are often observed in this region. The plasma sheet as a whole is reported to be thinnest near the midnight region and is about twice as thick near the flanks of the magnetotail. The plasma energy density is comparable to the magnetic field energy density in the plasma sheet region. In the middle of this region lies the neutral sheet where the magnetic field is weak. The field reverses its direction from pointing sunward to pointing tailward or vice versa as the neutral sheet is crossed. The region of the neutral sheet is considered to be an ideal site for magnetic reconnection. Magnetic reconnection is a prime candidate to account for many impulsive energy release phenomena observed in the magnetotail.

Schindler (1974) and Hones (1977) developed theories regarding the macroscopic effects of reconnection upon the magnetic topology of the magnetotail. In particular, they differentiated between the effects of reconnection involving closed magnetic field lines constituting the plasma sheet as opposed to the open flux comprising the lobes. In two dimensions, the formation of one or more X–lines on closed field lines always leads to the formation of magnetic loops or islands (Schindler, 1974). Similarly, reconnection at a single X–line involving the oppositely directed lobe field lines creates closed and interplanetary field lines on the Earthward and tailward sides of the neutral line, respectively. Hones (1977) first proposed a sequence of events by which closed field line reconnection in the plasma sheet is followed by open flux reconnection to produce substorms. Most notably, he predicted the formation of closed loops of magnetic flux between the near and distant neutral lines

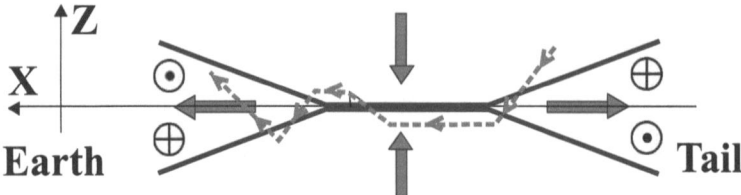

Figure 2.3: *Sketch of Cluster observations of Hall magnetic fields and current sheet structure around the reconnection region; the red dashed line indicates the trajectory of Cluster (Courtesy of A. Runov).*

which he called plasmoids. Such plasmoids were observed in the mid–1980's by ISEE 3 (e.g., Hones et al., 1984; Moldwin and Hughes, 1992). Intensive investigation was then undertaken using the more comprehensive instrumentation carried by Geotail (e.g., Ieda et al., 1998; Slavin et al., 2003). Sergeev et al. (1992) used different spacecraft and ground–based observations to study minute–scale features which may be considered as the main building blocks in the dissipative process responsible for substorms. The observed signatures consist of a high–speed flow burst with an accompanying increase of the normal magnetic field component, which are referred to as Nightside Flux Transfer Events (NFTEs), since they resemble similar features as FTEs. Such structures have also been studied by Sergeev et al. (1987; 1992) and were interpreted in terms of a transient reconnection model.

Since the Cluster mission, four spacecraft observations enable us to obtain spatial gradient of different parameters continuously. For example, the current density obtained from the gradient in the magnetic field is an essential parameter for magnetotail diagnostics. Using the gradient information obtained from the four Cluster spacecraft during current sheet crossings, current sheet structures in the vicinity of X-line were reconstructed by Runov et al. (2003) as illustrated in Fig. 2.3. A thin current sheet with a half-thickness of about one ion gyro radius was found for the crossing closest to the X-line, whereas the outer crossings showed bi-furcated current sheet profiles. Changes in the curvature of the field for the different current sheet crossings illustrated in the figure were consistent with a X-line motion from Earthward and tailward of the spacecraft. Furthermore, consistent feature of the field disturbance associated with Hall-current at both side of X-line were identified, confirming that the spacecraft traversed the ion diffusion region.

The importance of multi-scale processes from heavy ions to electron kinetics in the reconnection processes have been identified by several Cluster observations. During thin current sheet intervals associated with crossings of the X-line during storm-time substorms, the O^+ pressure and density were observed to be dominated compared to those of H^+ (Kistler et al., 2005). In such current sheets, the O^+ were observed to execute Speiser-type serpentine orbits across the tail and were found to carry about 5-10% of the cross-tail current (Kistler et al., 2005). The O^+ in the reconnection region was suggested to experience a ballistic acceleration (Wygant et al., 2005) based on the observation of a large amplitude bipolar electric field. These large-amplitude (up to 50 mV/m) solitary waves, identified as electron holes, have been observed during several plasma sheet encounters that have been identified as the passage of a magnetotail reconnection X-line (Cattell et al., 2005). The electron holes were seen near the outer edge of the plasma sheet, within and on the edge of a density cavity, at distances in the order of a few ion inertial lengths from the center of the current sheet. Based on a detailed comparison with simulations, Cattell et al. (2005) suggested that the observed nonlinear wave mode, electron holes, may play an important role in reconnection by scattering and energizing electrons.

Whereas in situ observations of ion diffusion region depend on the rare chance that the spacecraft being located at the right place at the right time (Øieroset et al., 2001; 2002; Borg et al., 2005), different effects of reconnection can be detected remotely from the reconnection site and can still contain useful information on temporal and spatial characteristics of reconnection. For example, the limited area of the Hall currents flowing in the ion diffusion region indicates that the current has to be closed including regions outside of the ion diffusion region. At the lobe side, the closure of the Hall current takes place via cold electron flowing into the ion diffusion region. At the outflow region, on the other hand, the accelerated electrons can carry the current into the ion diffusion region. Cluster observed such field-aligned electrons related to the Hall current system consistent with previous observations, but also obtained fine structures indicating multiple temporal or spatial properties of reconnection (e.g., Nakamura et al., 2004). Fig. 2.4 illustrates multipoint observations by Cluster in the northern hemisphere and by Geotail in the southern hemisphere detecting these field aligned currents possibly connected to the ion diffusion region from (Nakamura et al., 2004). Cluster four space also allowed to determine the scale size of these field aligned currents which suggested that the scale size of the downward current was at maximum comparable to the ion inertia length so that it plausibly connects to the near-Earth X-line and is driven by Hall effects in the reconnection region. This interhemispheric observation supported the theoretical prediction that the downward current region is thin because on the lobe side, the ions may travel a substantial fraction of the ion inertia length until their motion

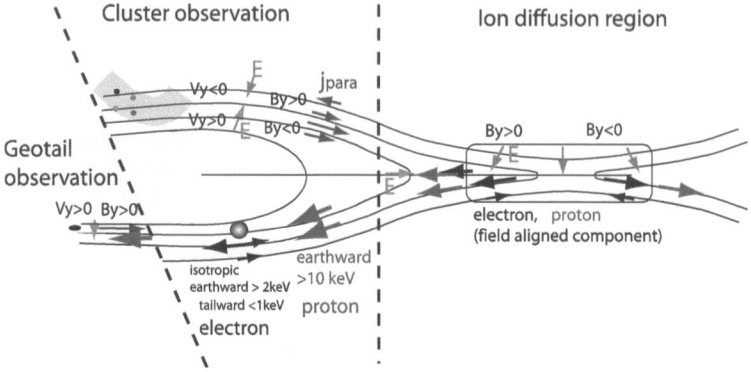

Figure 2.4: *Summary of observations by Cluster and Geotail during a transient entry into the plasma sheet during a substorm event and illustration of the possible relationship to the reconnection region. For Cluster observations field signatures are showing in the northern hemisphere, while particle signatures are illustrated in the southern hemisphere; adapted from Nakamura et al., 2005.*

separates from that of the electrons.

Using multi-composition plasma observation by Cluster, slow-mode shocks connected to the ion diffusion region have been analyzed by Eriksson et al. (2004) taking into account also the contribution from the oxygen during a substorm X–line event when Cluster observed fast tailward and Earthward flows. The successful joint Walén and slow shock analysis on the tailward flows within the plasma sheet presented further evidence in favor of Petschek-type reconnection at distances $X_{GSM} > $ -19 R_E of the near-Earth magnetotail.

Cluster succeeded to obtain detailed characteristics of earthward propagating southward then northward magnetic field disturbances related to plasmoids/flux rope (Slavin et al., 2003; Zong et al., 2004), travelling compression region (Owen et al., 2005) and nightside flux transfer events (Sergeev et al., 2005; Semenov et al., 2005a; Penz et al., 2006a). Multi-point analysis by Cluster were used to measure the current density and check the force-free model and energetic particle boundaries to show the structures of the plasmoid/flux rope. Yet, since the plasma flow jetting toward the Earth are significantly influenced by the strong dipolar field and pressure gradient, it still remains unknown to what extent these structures can be treated as motions of a stable structure in the analysis. As described in the following, similar magnetic features were rather interpreted as transient profiles of a remote X–line due to its change in the reconnection rate and used to determine the location of the X–line (Sergeev et al., 2005; Semenov et al., 2005a; Penz et al., 2006a).

2.6 Global picture of reconnection in the Earth's magnetosphere

If a front of southward orientated magnetic field approaches the magnetosphere, the magnetic field lines are stretched, and the magnetic field intensity increases. Thus, a current sheet develops at the magnetopause and the amount of free energy increases. Magnetic reconnection may start at any point at the dayside magnetopause, however, it is most likely that it will set on somewhere near the subsolar point where the field line stretching and hence the magnetopause current intensity are a maximum (Fig. 2.5). After reconnection has started, the fronts of the FR regions move towards the cusps, and the flux of the reconnected magnetic field increases. Beyond the cusps, the magnetopause currents change their direction, therefore the Ampère forces are directed sunward so that they hinder the plasma motion. This hindrance is compensated by the gradient of the plasma pressure which arises as the result of the plasma acceleration and heating in the course of magnetic reconnection. Thus, at the high latitude magnetopause beyond the cusps, the plasma is moving against the Ampère forces, performing some work. As a result, a corresponding amount of the electromagnetic energy produced is transported into the magnetotail in the form of the Poynting vector. The direction of the Poynting vector gives the direction of the $\mathbf{E} \times \mathbf{B}$ drift motion, which implies that the reconnected field lines from both hemispheres are convected into the equatorial plane, whereby the so–called neutral sheet current develops to separate these antiparallel lobe field lines. The first of these so–called open magnetospheres was developed by Dungey (1961).

Under steady–state conditions, the driving electric field remains switched on during the whole process, so that more and more field lines get reconnected on the dayside, convect back into the tail and consequently more and more magnetic flux is accumulated in the neutral sheet. Thus, dayside reconnection provokes a piling up process in the tail region, which will be limited by reconnection in the nightside region. Hence, dayside merging adds magnetic flux to the open field line region and nightside merging subtracts from it. Over a long enough period, the total flux of open field lines does not change and therefore, the dayside and nightside merging rate must be equal on the average. In summary, one can say that the solar wind–magnetosphere

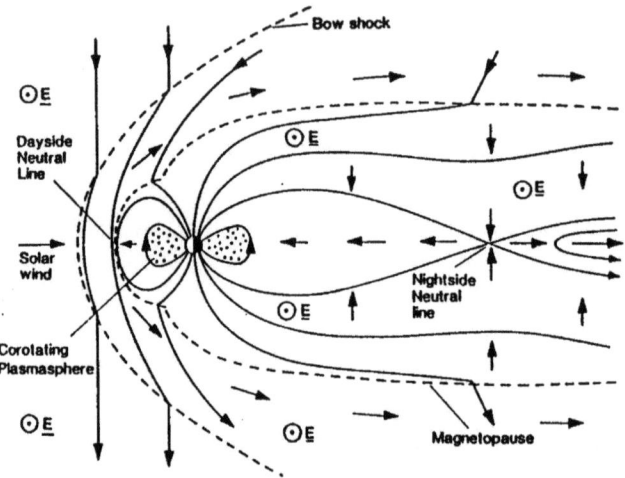

Figure 2.5: *Global pattern of plasma transport in an open magnetosphere (Saunders, 1991).*

interaction may be represented by a chain of energy transformations:

$$\text{solar wind kinetic energy}$$
$$\Downarrow$$
$$\text{magnetic energy (pile–up process)}$$
$$\Downarrow$$
$$\text{kinetic and thermal energy (dayside reconnection)}$$
$$\Downarrow$$
$$\text{electromagnetic energy (Poynting vector)}$$
$$\Downarrow$$
$$\text{magnetic energy (pile–up process)}$$
$$\Downarrow$$
$$\text{kinetic and thermal energy (nightside reconnection)}.$$

Concerning the plasma motion one can say that the plasma is accelerated by dayside reconnection, leaves the reconnection site in the form of high speed streams, reaches the magnetospheric lobe region, where the dawn to dusk electric field forces it to move again down to the equatorial plane, and finally nightside reconnection pushes the plasma back into the inner magnetosphere (Fig. 2.5).

This processes can be related with the different stages of a substorm. During the initial phase (growth phase) of a substorm, all energy entering the magnetotail lobes is transformed into magnetic energy ($\partial B/\partial t > 0$). However, the intensity of the magnetotail magnetic field and the corresponding current density in the plasma sheet cannot increase indefinitely, and when the latter approaches some threshold value, the process of magnetic reconnection is initiated in the magnetotail and FR regions develop. This moment may be considered as the onset of the expansion phase of the substorm. In the course of time, the FR regions are detached from the reconnection site and propagate both Earth– and tailward. In the Earthward direction, the reconnection–associated currents are propagated along the neutral sheet until they reach the edge of the current sheet. At this boundary, the currents are deflected into the ionosphere along the so–called Birkeland currents. During this phase $\partial B/\partial t < 0$, and the energy stored in the magnetotail lobes is spent to accelerate and heat the plasma within the plasma sheet and to generate intense current systems in the auroral magnetosphere and ionosphere. Finally, when the free magnetic energy in the lobes is exhausted, the development of magnetospheric disturbances enters a steady–state regime, and $\partial B/\partial t \approx 0$. At this stage of the disturbance, the energy which enters the magnetosphere is transformed directly into energy of the magnetospheric plasma and the auroral current system

3 MHD equations and magnetic reconnection models

There are two possible methods for analyzing plasma behavior: *a*) studying the effects of the electric and magnetic fields on test particles, or *b*) a statistical approach, in which we model the plasma as a collective. Nevertheless, these are not two strictly separated methods, rather they can be considered as the two endpoints of a continuous spectra with different models using different aspects of these approaches.

The former method is only strictly valid for very low particle number densities since it can then be assumed that the particle motions do not affect the fields. Nevertheless, this method can provide us with a wealth of information, especially regarding single–particle dynamics with respect to many different electric and magnetic field configurations. For the remainder of the present work we shall consider only the statistical approach. Here we treat groups of particles as entities and assume the effects of inter–particle interaction to be negligible compared to the effects of the medium. Since the plasma appears charge–neutral on macroscopic length scales, we can ignore the identities of single particles and consider the plasma as a continuum. Thus, there are strong parallels with fluid mechanics, but the introduction of electromagnetic forces and effects arising from the motions of charged particles within the plasma give rise to a far richer behavior and more diverse range of phenomena. The resulting theory developed to describe such an electromagnetic fluid is called magnetohydrodynamics (MHD).

3.1 The basic equations

3.1.1 The equation of continuity

If we consider a streaming fluid, the mass streaming through an arbitrary area can be written as $\rho(\mathbf{r},t)\,\mathbf{v}(\mathbf{r},t)\cdot\mathbf{n}\,da$, where $\rho(\mathbf{r},t)$ is the mass density, $\mathbf{v}(\mathbf{r},t)$ is the velocity and $\mathbf{n}$ is the unit outward normal vector at a certain area element da. Integration over a closed surface leads to

$$\oint \rho \mathbf{v}\cdot\mathbf{n}\,da = \int \nabla\cdot(\rho\mathbf{v})\,d^3x \equiv -\frac{\partial}{\partial t}\int \rho\,d^3x = -\int \frac{\partial \rho}{\partial t}\,d^3x\,.$$

Here we use the divergence theorem

$$\int_V \nabla\cdot\mathbf{A}\,d^3x = \oint_S \mathbf{A}\cdot\mathbf{n}\,da\,, \tag{3.1}$$

and demand that the flux through a surface must be equivalent to the change of mass inside the volume bounded by the surface. Thus, it can be referred to as mass conservation. Since the surface is arbitrary, the arguments of the integrals must be the same. This gives the equation of continuity

$$\frac{\partial \rho}{\partial t} + \nabla \cdot (\rho \mathbf{v}) = 0. \qquad (3.2)$$

3.1.2 The equation of motion

Using Newton's equation $\mathbf{F} = m\mathbf{a}$ and dividing by the volume V, we can write

$$\mathbf{k} = \rho \frac{d\mathbf{v}}{dt} = \rho \left(\frac{\partial \mathbf{v}}{\partial t} + (\mathbf{v} \cdot \nabla) \mathbf{v} \right),$$

where $\mathbf{k}$ is called force density. We can introduce different forces and write a general equation

$$\rho \left(\frac{\partial \mathbf{v}}{\partial t} + (\mathbf{v} \cdot \nabla) \mathbf{v} \right) = \frac{1}{c} \mathbf{j} \times \mathbf{B} - \nabla p + \nabla \mathbf{S} + \mathbf{f}_g, \qquad (3.3)$$

where p is the thermal pressure, $\mathbf{j}$ is the current density, $\mathbf{S}$ is the viscous stress tensor and $\mathbf{f}_g$ is an external force density. In our case, we neglect the electric field (because MHD assumes a quasi–neutral plasma), the stress tensor and all external forces. So we get the equation of motion as

$$\rho \left(\frac{\partial \mathbf{v}}{\partial t} + (\mathbf{v} \cdot \nabla) \mathbf{v} \right) - \frac{1}{c} \mathbf{j} \times \mathbf{B} - \nabla p. \qquad (3.4)$$

3.1.3 The adiabatic law

For ideal MHD, we assume that the particles are in a local thermodynamic equilibrium and we can demand that there are variables of state, which obey the laws of thermodynamic. In particular, a thermal equation of state must be fulfilled

$$p = p(\rho, T).$$

Now we can write that $pV^\gamma = const$, where $\gamma = c_p/c_v$ is the adiabatic coefficient. The adiabatic law follows as

$$\frac{d}{dt} \left(\frac{p}{\rho^\gamma} \right) = 0. \qquad (3.5)$$

3.1.4 The Maxwell equations and Ohm's law

For our purpose we use the Maxwell equations in the following form:

Faraday's law: $\quad \nabla \times \mathbf{E} = -\frac{1}{c} \frac{\partial \mathbf{B}}{\partial t}, \qquad (3.6)$

Ampère's law: $\nabla \times \mathbf{B} = \dfrac{4\pi}{c}\mathbf{j} + \dfrac{1}{c}\dfrac{\partial \mathbf{E}}{\partial t}$, (3.7)

Coulomb's law: $\nabla \cdot \mathbf{E} = 4\pi \rho_e$, (3.8)

Absence of free magnetic poles: $\nabla \cdot \mathbf{B} = 0$. (3.9)

In the low-frequency case ($\omega \ll \omega_p, \omega_c$), for the Maxwell current is much smaller than $4\pi/c\,\mathbf{j}$,

$$\frac{1}{c}\frac{\partial \mathbf{E}}{\partial t} \ll \frac{4\pi}{c}\mathbf{j}.$$

Therefore, Equation (3.7) can be written as

$$\nabla \times \mathbf{B} = \frac{4\pi}{c}\mathbf{j}$$

and taking the divergence gives

$$\nabla \cdot \mathbf{j} = 0. \qquad (3.10)$$

Finally, we use Ohm's law in the form

$$\mathbf{j} = \sigma\left(\mathbf{E} + \frac{1}{c}\mathbf{v}\times\mathbf{B}\right), \qquad (3.11)$$

where σ denotes the scalar electrical conductivity.
There are two possibilities to deal with Equation (3.9):

- If the problem is time–dependent, we can take the divergence of Equation (3.7), giving $0 = -1/c\,\partial/\partial t(\nabla \cdot \mathbf{B})$. Using $\nabla \cdot \mathbf{B} = 0$ for $t = 0$, we do not need Equation (3.9) anymore.

- If the problem is not time–dependent, Equation (3.6) gives $\nabla \times \mathbf{E} = 0$ and we can introduce a potential $\mathbf{E} = -\nabla\phi$. In this case we need Equation (3.9), since substituting $\mathbf{E}$ with ϕ eliminates two components, but also Equation (3.7) is eliminated.

3.1.5 Simplification of the system of equations

- Elimination of the space–charge density:
Since we are considering characteristic length scales much larger than the Debye length, we can assume quasi–neutrality:

$$\frac{n_i - n_e}{n_e} \ll 1 \Rightarrow \rho_e \approx 0.$$

- Elimination of the current:
  Substituting Equation (3.7) into the equation of motion gives

  $$\rho \frac{d\mathbf{v}}{dt} = -\frac{1}{4\pi} \left(\mathbf{B} \times (\nabla \times \mathbf{B}) \right) - \nabla p.$$

  Substituting Ohm's law into Ampère's law leads to

  $$\nabla \times \mathbf{B} = \frac{4\pi}{c} \sigma \left(\mathbf{E} + \frac{1}{c} \mathbf{v} \times \mathbf{B} \right),$$

  so that $\mathbf{j}$ is eliminated in the system of equations.

- Homogeneous conductivity:
  We can take the curl of the previous equation and use Faraday's law to get

  $$\frac{\partial \mathbf{B}}{\partial t} = \frac{c^2}{4\pi\sigma} \Delta \mathbf{B} + \nabla \times (\mathbf{v} \times \mathbf{B}), \qquad (3.12)$$

  so that the electric field $\mathbf{E}$ is eliminated.

Considering the limiting cases of Equation (3.12) leads to:

$$\frac{\partial \mathbf{B}}{\partial t} = \nabla \times (\mathbf{v} \times \mathbf{B}), \qquad \sigma \to \infty \qquad (3.13)$$

$$\frac{\partial \mathbf{B}}{\partial t} = \frac{c^2}{4\pi\sigma} \Delta \mathbf{B}, \qquad \mathbf{v} \to 0. \qquad (3.14)$$

The first of these equations is used to show flux freezing in an ideal plasma, while the second one is simply a diffusion equation. The diffusion coefficient $D = c^2/(4\pi\sigma)$ in the second equation is indirect proportional to the magnetic Reynolds number R_m according to $D \sim L/R_m$, where L is a characteristic length scale. Being R_m generally very large for space plasmas, the frozen-in-flux condition holds to a very high accuracy and D will be negligible. In this case, the second term can be neglected and hence ideal MHD can be applied.

3.1.6 The reduced system of MHD

Collecting previous results gives the following system of equations for an infinitely conducting media:

$$\rho \frac{\partial \mathbf{v}}{\partial t} + \rho \left(\mathbf{v} \cdot \nabla \right) \mathbf{v} = -\nabla \left(p + \frac{B^2}{8\pi} \right) + (\mathbf{B} \cdot \nabla) \frac{\mathbf{B}}{4\pi}, \qquad (3.15)$$

$$\frac{\partial \mathbf{B}}{\partial t} = \nabla \times (\mathbf{v} \times \mathbf{B}), \qquad (3.16)$$

$$\frac{\partial \rho}{\partial t} + \nabla \cdot (\rho \mathbf{v}) = 0, \tag{3.17}$$

$$\nabla \cdot \mathbf{B} = 0, \tag{3.18}$$

$$\frac{d}{dt}\left(\frac{p}{\rho^\gamma}\right) = 0. \tag{3.19}$$

These basic equations are essential to describe different models of magnetic reconnection. However, it should be noted that the process of magnetic reconnection itself cannot be described by using ideal MHD. According to the ideal induction equation (Eq. 3.13), field lines move with the plasma, and the flux is frozen into the plasma (Alfvén, 1943). Therefore, two interacting plasmas cannot diffuse into each other and a thin current sheet will develop, across which the magnetic field topology changes. However, in a narrow current sheet, like the Earth's magnetopause, there can be regions where the characteristic length scales L are getting small. There, the diffusion coefficient D becomes larger and magnetic diffusion becomes important, so that the diffusion term in Eq. 3.12 cannot be neglected anymore. Thus, a region where ideal MHD is not valid anymore will develop. The previously separated magnetic fields loose their identity in this diffusion region and can merge or reconnect. Therefore, different models where developed in order to introduce a diffusion region and allow for magnetic reconnection.

3.2 Sweet–Parker model

The Sweet–Parker model (see e. g., Parker, 1957; Sweet, 1958; Priest and Forbes, 2000) consists of a simple diffusion region of length $2L$ and width $2l$ lying between oppositely directed magnetic fields (Figure 3.1). Our aim is to get some information about the velocity v_i of the plasma, when it is entering the diffusion layer. We make just an order–of–magnitude analysis for simplicity. Since the electric field is uniform for a steady two–dimensional state, we can evaluate Ohm's law (3.11) in the inflow region, where the current vanishes, as

$$E = \frac{1}{c} v_i B_i, \tag{3.20}$$

while Ohm's law gives for the center of the diffusion region, where the magnetic field vanishes,

$$E = \frac{j}{\sigma}. \tag{3.21}$$

Using Ampère's law (3.7) in the center of the diffusion region gives

$$j \approx \frac{c}{4\pi} \frac{B_i}{l}. \tag{3.22}$$

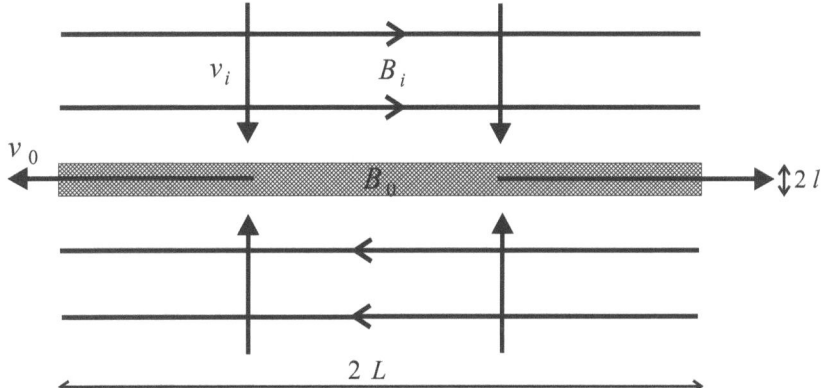

Figure 3.1: *Sweet–Parker model of reconnection. The diffusion region is the shaded area, the plasma velocities are indicated by thick–headed arrows, and the magnetic field lines by thin–headed arrows.*

Eliminating E and j between these three equations gives for the inflow plasma velocity

$$v_i = \frac{\eta}{l}, \tag{3.23}$$

where $\eta = c^2/(4\pi\sigma)$ is the magnetic diffusivity.

Conservation of mass requires that the rate ($4\rho L v_i$) at which mass is entering the diffusion region from both sides must be the same as the the rate ($4\rho l v_0$) at which it is leaving on both ends, so that

$$L v_i = l v_0. \tag{3.24}$$

Now we can eliminate the width l to get the inflow velocity as

$$v_i = \sqrt{\frac{\eta v_0}{L}}. \tag{3.25}$$

To determine the outflow plasma velocity we use the equation of motion (3.4). From Equation (3.22) we get the electric current, and so the Lorentz force along the current sheet is $(\mathbf{j} \times \mathbf{B})_x \approx j B_0 = c B_i B_0 / (4\pi l)$. This force accelerates the plasma from rest at the neutral point to v_0 over a distance L. So the equation of motion leads to

$$\rho \frac{v_0^2}{L} \approx \frac{B_i B_0}{4\pi l}. \tag{3.26}$$

Using $\nabla \cdot \mathbf{B} = 0$ gives for the outflow velocity

$$v_0 = \frac{B_i}{\sqrt{4\pi\rho}} \equiv v_A. \quad (3.27)$$

Not surprisingly, we have found that the magnetic force accelerates the plasma to Alfvén speed. The fields therefore reconnect for this basic model at a speed given by Equation (3.25) as

$$v_i = \frac{v_A}{\sqrt{R_m}}, \quad (3.28)$$

where the magnetic Reynolds number is given as

$$R_m = Lv_A/\eta. \quad (3.29)$$

Usually, in space plasmas the magnetic Reynolds number is much larger than unity, so that

$$v_i << v_A, \ B_0 << B_i. \quad (3.30)$$

For example, in stellar coronae R_m lies between 10^6 and 10^{12} and so the fields reconnect at between 10^{-3} and 10^{-6} of the Alfvén speed. This is much too slow to describe phenomena like stellar flares or coronal heating. Therefore, other models have been developed to describe these processes more appropriate.

3.3 Petschek–mechanism for magnetic reconnection

Reconnection is said to be fast, when the reconnection rate M_e is much larger than the Sweet-Parker rate (Equation 3.28). These fast regimes of reconnection contain a tiny Sweet–Parker diffusion region as shown in Figure 3.2. Petschek (1964) realized that a slow magnetoacoustic shock (for details see Priest and Forbes, 2000) provides another way of converting magnetic energy into heat and kinetic energy. In the switch–off limit, the shock propagates at a speed $v_s = B_N/\sqrt{4\pi\rho}$ into a medium at rest and has the effect of turning the magnetic field towards the normal and therefore decreasing the downstream field strength. At the same time, the shock accelerates the plasma to the Alfvén speed parallel to the shock front. Now, if the upstream plasma is moving downwards at the same speed (v_s) as the shock is trying to propagate upwards, then the shock front will remain stationary.

Petschek appreciated that the Sweet–Parker diffusion region would act as a source for four slow magnetoacoustic shocks, which propagate in different directions from the diffusion region and which stand in the flow when a steady state is reached (Figure 3.3 a). The creation of standing shock waves without any solid obstacle in the flow has been confirmed by numerous numerical

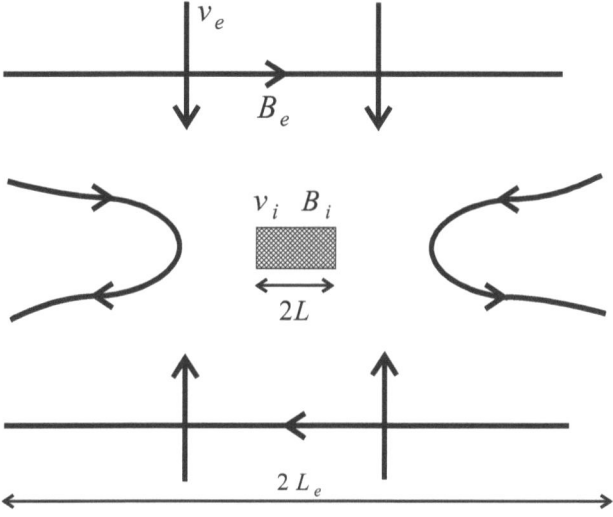

Figure 3.2: *Notation for fast reconnection.*

experiments (e. g., Scholer, 1989).

The Petschek analysis is disarmingly simple. The magnetic field decreases substantially from an uniform value (B_e) at large distances to a value B_i at the entrance to the diffusion region, while the flow speed increases from v_e to v_i. The effect of the shocks is to provide a normal magnetic field component (B_N), which is associated with the distortion in the inflow field from the uniform value at large distances. Thus, if the inflow field is potential, the distortion may be regarded as being produced by a series of monopole sources along the x–axis between $-L_e$ and $-L$ and between L and L_e (Figure 3.3 b).

The inflow region therefore consists of slightly curved field lines. The magnetic field there is the sum of a uniform horizontal field and the field obtained by solving Laplace's equation with the boundary conditions, that the magnetic field vanishes at large distances and that the normal component of the field be B_N along the shock waves and vanishes at the diffusion region. To lowest order, the inclination of the shocks may be neglected, and so the problem is to find a solution in the upper half–plane, which vanishes at infinity and which is equal to $2B_N$ between L and L_e on the x–axis and, by symmetry, $-2B_N$ between $-L_e$ and $-L$.

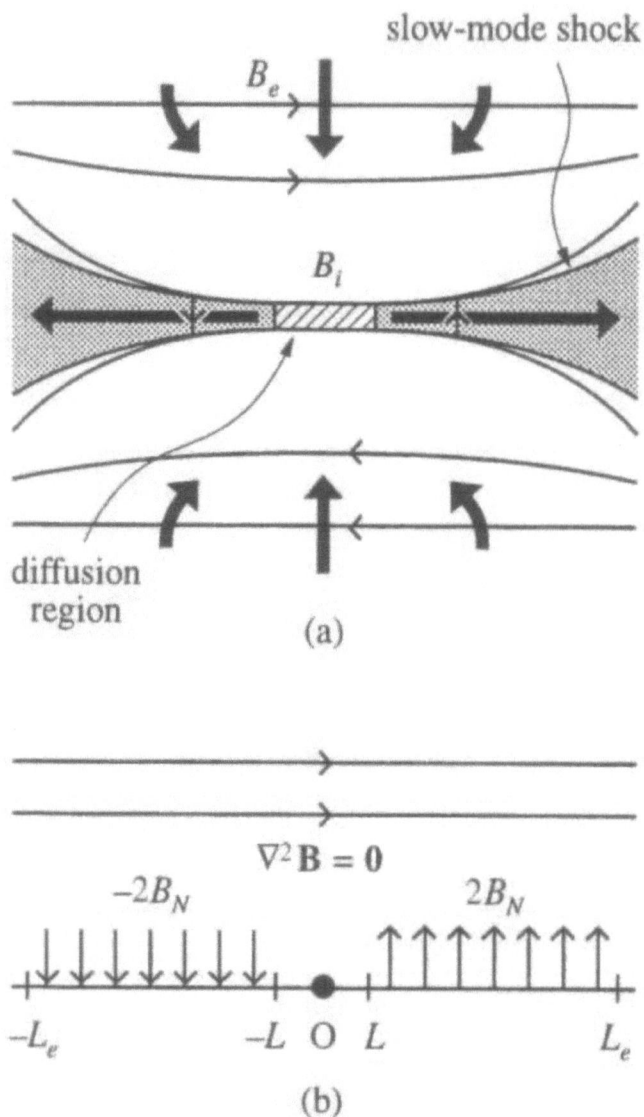

Figure 3.3: *(a) Petschek's model, in which the central shaded region is the diffusion region and the other two shaded regions represent plasma that is heated and accelerated by the shocks. (b) Notation for the analysis of the inflow region (from Priest and Forbes, 2000).*

Now we may regard the normal component on the x-axis as being produced by a continuous series of poles. If each pole produces a field m/r at a distance r, then the flux produced in the upper half-plane by that pole will be πm; but if the pole occupies a length dx of the x-axis, the flux is also $2B_N dx$, so that $m = 2B_N/\pi$. Then, integrating along the x-axis gives the field at the origin produced by the poles as

$$\frac{1}{\pi} \int_{-L_e}^{-L} \frac{2B_N}{x} dx - \frac{1}{\pi} \int_{L}^{L_e} \frac{2B_N}{x} dx. \qquad (3.31)$$

Adding this to the field at infinity gives the diffusion region inflow field as

$$B_i = B_e - \frac{4B_N}{\pi} \log \frac{L_e}{L}. \qquad (3.32)$$

Since slow shocks travel with Alfvén speed in the switch-off limit, which will be explained in the next section in more detail, we can rewrite Equation (3.32) as

$$B_i = B_e \left(1 - \frac{4M_e}{\pi} \log \frac{L_e}{L}\right), \qquad (3.33)$$

where $M_e = v_e/v_{Ae}$ is the external reconnection rate. Petschek suggested, that the mechanism chokes itself off when B_i becomes too small, and so he estimated a maximum reconnection rate M_e^* by putting $B_i - B_e/2$ in Equation (3.33) to give

$$M_e^* \approx \frac{\pi}{8 \log R_{me}}. \qquad (3.34)$$

In value this lies typically in the range of 0.1 since $\log R_{me}$ is slowly varying, so we see that for typical R_{me} values the reconnection is much faster than the Sweet–Parker rate. Thus, for twenty years the problem of fast reconnection was thought to have been solved completely by Petschek, until in the 1980s a new generation of reconnection solutions were discovered, with Petschek's mechanism as a special case (Priest and Forbes, 1986). It is now realized to be one member of a whole family of fast reconnection regimes (Priest and Forbes, 2000).

4 Direct model of time–dependent Petschek–type reconnection in an incompressible plasma

The aim of this work is the development of an inverse model of Petschek–type magnetic reconnection. In an inverse model, the reconnection electric field can be calculated from magnetic field disturbances caused by magnetic reconnection. However, in order to establish an inverse model, first it is necessary to develop a direct model. In a direct model, the magnetic field perturbations caused by a given reconnection electric field can be calculated. In this section, we derive an analytical expression for the MHD quantities. Based on this expressions, we will perform an inversion in the next section in order to solve the inverse problem.

4.1 Basic configuration

For our purpose, we consider oppositely directed magnetic fields, which are separated by an infinitely thin tangential discontinuity. Across a tangential discontinuity, the normal components of the magnetic field and plasma velocity are zero ($B_n = 0$, $v_n = 0$). To fulfill the Rankine–Hugoniot relations, the density and the tangential components of the magnetic field and velocity may change arbitrarily, subject only to the requirement that the total pressure, which is the sum of the thermal and the magnetic pressure, $p + B^2/(8\pi)$, stays constant (Heyn et al., 1988). Since there is no mass flow and no magnetic connection across a tangential discontinuity, there will be no electric field component along it ($E_t = 0$). If we assume that the plasma conductivity decreases in a certain region, the generation of a reconnection electric field, and therefore a normal component of the magnetic field will follow (Biernat et al., 1987). Any local deviation of the perfect conductivity approximation due to the reconnection electric field will cause a reconfiguration of the tangential discontinuity. The surface gets non–linearly unstable and decays into a system of MHD wave modes.

To simplify the problem, we introduce the restriction of so–called weak reconnection (Petschek, 1964) which implies, that the reconnection electric field $E(t)$ is much smaller than the Alfvén electric field $E_A = v_A B_0/c$, where B_0 and $v_A = B_0/\sqrt{4\pi\rho}$ are the initial magnetic field and the Alfvén velocity, respectively. Within this restriction, we are considering small components of the magnetic field and velocity in a direction normal to the tangential discontinuity:

$$B_n^2 << B_t^2, \qquad (4.1)$$

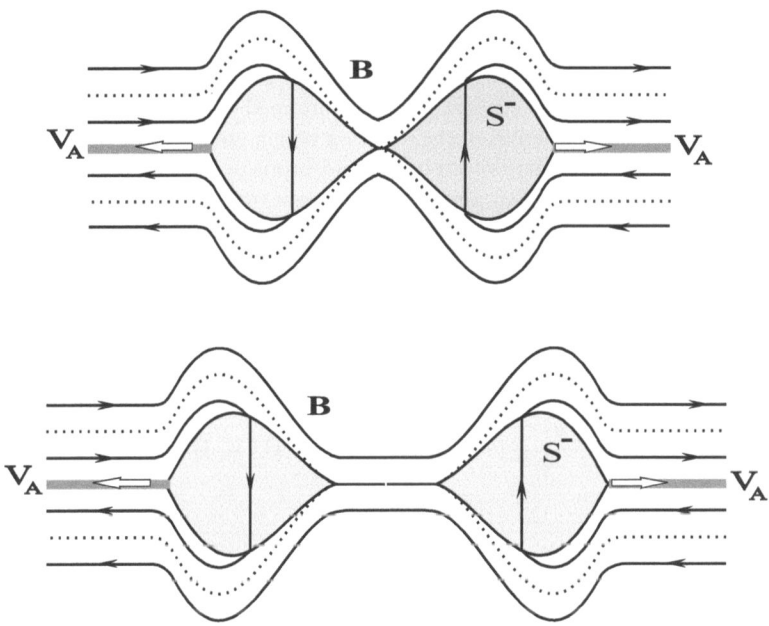

Figure 4.1: *Basic configuration for time–dependent Petschek–type magnetic reconnection. In the upper panel, the reconnection electric field is still active, implying that the outflow regions with heated and accelerated plasma (shaded areas) are still attached to the diffusion region. In incompressible plasma, the outflow region is bounded by a so–called Petschek shock (S^-), which is a merger of a slow shock and an Alfvén wave. In the lower panel, the reconnection electric field ceased, leading to a detachment of the outflow regions, which continue to propagate along the current sheet (modified from Semenov et al. (2004b).*

$$\rho v_n^2 \ll p + \frac{B_t^2}{8\pi}. \tag{4.2}$$

Thus, to lowest order, the total pressure is conserved across the discontinuity. As a result, there is no pressure gradient that could drive a strong fast shock. Further, in the incompressible case $\rho = const$, which is considered in the following, the Alfvén wave merges with the slow shock, forming a so–called Petschek shock. These Petschek shocks bound a region of accelerated and heated plasma, the field reversal or outflow region (Fig. 4.1). After some time, the reconnection electric field has dropped to zero and the outflow regions become detached from the diffusion region and continue to propagate along the tangential discontinuity. These propagating shocks are causing disturbances in the ambient plasma environment, which can be measured by a satellite. In the following, we will show how these measured disturbances can be calculated by using a time–dependent Petschek–type model of transient magnetic reconnection.

To study the problem of symmetric reconnection in an incompressible plasma ($\rho = const$) away from the reconnection line, the ideal magnetohydrodynamic (MHD) equations in normalized form (e.g., Semenov et al., 2005b)

$$\frac{\partial \mathbf{v}}{\partial t} + (\mathbf{v} \cdot \nabla)\mathbf{v} = -\nabla P + (\mathbf{B} \cdot \nabla)\mathbf{B}, \tag{4.3}$$

$$\frac{\partial \mathbf{B}}{\partial t} = \nabla \times (\mathbf{v} \times \mathbf{B}), \tag{4.4}$$

$$\nabla \cdot \mathbf{v} = 0, \tag{4.5}$$

$$\nabla \cdot \mathbf{B} = 0, \tag{4.6}$$

can be used. Here, $\mathbf{B}$ denotes the magnetic field, and $\mathbf{v}$ is the bulk velocity of the plasma. The background magnetic field $\mathbf{B}^{(0)}$ and the total pressure P are assumed to be constant. Additionally, we consider a fixed plasma, meaning that $\mathbf{v}^{(0)} = 0$ in the inflow region in zero order. The normalization is done with respect to the initial magnetic field $B^{(0)}$, the initial Alfvén velocity v_A, the time duration of a reconnection pulse $T^{(0)}$, the Alfvén electric field $E_A = v_A B^{(0)}/c$, and the length scale $v_A T^{(0)}$.

In the following, the direct method of time–dependent Petschek–type magnetic reconnection developed by Heyn and Semenov (1996) and Semenov et al. (2004a) is presented. The direct problem is based on the Cagniard-deHoop method (Lamb, 1904; Cagniard, 1939; deHoop, 1960; Heyn and Semenov, 1996), which allows to calculate the disturbances in the ambient plasma environment for a given reconnection electric field. Using this method, the magnetic field and velocity components are found to be convolution integrals of the reconnection rate and an integration kernel, which includes the magnetic field parameters and the distance information. We show that the plasma and magnetic field behavior is in agreement with the picture proposed for NFTEs.

4.2 Displacement vector in $\mathcal{L}$–$\mathcal{F}$ space

The starting point is the normalized equation of motion

$$\frac{\partial \mathbf{v}}{\partial t} + (\mathbf{v} \cdot \nabla)\, \mathbf{v} = -\nabla P + (\mathbf{B} \cdot \nabla)\, \mathbf{B}\,. \tag{4.7}$$

Using linear perturbation theory and assuming that $\mathbf{B}_0 = \mathbf{v}_0 = const$, where subscript 0 denotes unperturbed quantities, we get

$$\frac{\partial \mathbf{v}}{\partial t} + (\mathbf{v}_0 \cdot \nabla)\, \mathbf{v} = -\nabla P + (\mathbf{B}_0 \cdot \nabla)\, \mathbf{B}\,. \tag{4.8}$$

We introduce the Laplace transformation $\mathcal{L}$ as

$$\hat{f}(x,z,p) = \int_0^\infty f(x,z,t)\, e^{-pt} dt\,, \tag{4.9}$$

and the inverse Laplace transformation $\mathcal{L}^{-1}$ as

$$f(x,z,t) = \frac{1}{2\pi i} \int_{\sigma-i\infty}^{\sigma+i\infty} \hat{f}(x,z,p)\, e^{pt} dp\,, \tag{4.10}$$

where σ is large enough that $\hat{f}(x,z,p)$ is defined for $\Re(p) \geq \sigma$. The Fourier transformation $\mathcal{F}$ is defined as

$$\tilde{f}(k,z,t) = \int_{-\infty}^\infty f(x,z,t)\, e^{-ikx} dx\,, \tag{4.11}$$

while the inverse Fourier transformation $\mathcal{F}^{-1}$ is given as

$$f(x,z,t) = \frac{1}{2\pi} \int_{-\infty}^\infty \tilde{f}(k,z,t) e^{ikx} dk\,. \tag{4.12}$$

It is obvious that derivations with respect to time in $\mathcal{L}$–space give

$$\frac{\partial}{\partial t} f(x,z,t) \to p\, \hat{f}(x,z,p)\,, \tag{4.13}$$

while derivations with respect to x in $\mathcal{F}$–space are leading to

$$\frac{\partial}{\partial x} f(x,z,t) \to i\, k\, \tilde{f}(k,z,t)\,. \tag{4.14}$$

If we apply $\mathcal{L}$ and $\mathcal{F}$ on the z–component of the equation of motion (4.7), we get
$$(p + v_0\, i\, k)\, \hat{\tilde{\mathbf{v}}} = -\frac{\partial P}{\partial z} + B_0\, i\, k\, \hat{\tilde{\mathbf{B}}}. \tag{4.15}$$
For convenience, we do not write the hat and tilde in the following, but if it is necessary, we write functions in the real space as $f(x,t)$ and functions in $\mathcal{L}$–$\mathcal{F}$ space as $f(k,p)$.

Now we introduce a displacement vector $\boldsymbol{\xi}$, which is defined as
$$\mathbf{v} = \left(\frac{\partial}{\partial t} + \mathbf{v}_0 \cdot \nabla\right) \boldsymbol{\xi}, \tag{4.16}$$
and
$$\mathbf{B} = (\mathbf{B}_0 \cdot \nabla)\, \boldsymbol{\xi}. \tag{4.17}$$
Inserting Equations (4.16) and (4.17) in the transformed equation of motion (4.15) leads to
$$(p + v_0\, i\, k)^2\, \boldsymbol{\xi} = -\frac{\partial P}{\partial z} + (B_0\, i\, k)^2\, \boldsymbol{\xi}. \tag{4.18}$$
Assuming that $v_0 = 0$ gives for the z–component of Equation (4.18)
$$\epsilon\, \xi_z = -\frac{\partial P}{\partial z}, \tag{4.19}$$
with
$$\epsilon = p^2 + B_0^2 k^2. \tag{4.20}$$
The z–component of the displacement vector fulfills the Laplace equation (Semenov et al., 2005a)
$$\Delta \xi_z = 0, \tag{4.21}$$
which gives
$$\frac{\partial^2 \xi_z}{\partial z^2} + k^2 \xi_z = 0 \tag{4.22}$$
in $\mathcal{L}$–$\mathcal{F}$ space. Solutions of this equation can be written as
$$\xi_{z+} = c_+ e^{-|k|z} \quad \forall\ z > 0, \tag{4.23}$$
and
$$\xi_{z-} = c_- e^{|k|z} \quad \forall\ z < 0, \tag{4.24}$$
where subscript $+$ and $-$ denote the upper and the lower inflow region, respectively. Inserting Equations (4.23) and (4.24) into Equation (4.19) leads to
$$\frac{\partial P_\pm}{\partial z} = -\epsilon_\pm \xi_{z\pm} = -\epsilon_\pm c_\pm e^{\mp |k|z}. \tag{4.25}$$

Integration with respect to z yields

$$P_{\pm} = \pm \frac{\epsilon_{\pm} c_{\pm}}{|k|} e^{\mp |k| z}. \qquad (4.26)$$

Using the total pressure balance $P_+ = P_-$ gives at $z = 0$ the dispersion relation

$$\epsilon_+ c_+ + \epsilon_- c_- = 0. \qquad (4.27)$$

At $z = 0$, we can define a source term in $\mathcal{L}\text{-}\mathcal{F}$ space as

$$Q(k, p) = \xi_{z+} - \xi_{z-} = c_+ - c_-. \qquad (4.28)$$

Inserting this source term in Equation (4.23) and (4.24), respectively, gives the displacement vector in $\mathcal{L}\text{-}\mathcal{F}$ space as

$$\xi_{z\pm}(k, p) = \pm \frac{\epsilon_{\mp}}{\epsilon_+ + \epsilon_-} Q(k, p) \, e^{\mp |k| z}. \qquad (4.29)$$

4.3 Source term in $\mathcal{L}\text{-}\mathcal{F}$ space

Now we know the displacement vector in $\mathcal{L}\text{-}\mathcal{F}$ space, but we still have no expression for the source term in $\mathcal{L}\text{-}\mathcal{F}$ space. Therefore, we show in this subsection how to derive the source term in $\mathcal{L}\text{-}\mathcal{F}$ space from the expression in coordinate–time space.

The shape of the shocks in real space is given as

$$f_+(x, t) = \xi_{z+}(x, t) + \Phi_+(t - |x|/B_a) = \xi_i(x, t) + \Phi_i(t - |x|/B_a), \qquad (4.30)$$

in the upper half plane, and as

$$f_-(x, t) = \xi_{z-}(x, t) + \Phi_-(t + |x|/B_b) = \xi_i(x, t) + \Phi_i(t + |x|/B_b), \qquad (4.31)$$

in the lower half plane, where B_a and B_b is the magnetic field in the upper and the lower half plane, subscript i denotes quantities in the outflow region, and $\xi_i(x, t)$ is the displacement vector in the outflow region. Additionally, we use

$$\Phi_k = \frac{1}{B_k} \int_0^t E(\tau) d\tau, \qquad (4.32)$$

where $E(t)$ is the reconnection electric field. Substracting Equation (4.31) from Equation (4.30), we get the source term in real space as

$$Q(x, t) = \Phi_i(t - |x|/B_a) - \Phi_i(t + |x|/B_b) + \Phi_-(t + |x|/B_b) - \Phi_+(t - |x|/B_a). \qquad (4.33)$$

Our aim is now to transform this source term in the $\mathcal{L}$–$\mathcal{F}$ space. Therefore, we apply $\mathcal{L}$ and $\mathcal{F}$ to the first term on the right hand side of Equation (4.33) yielding

$$Q^{(1)}(p,k) = \int_0^\infty dt\, e^{-pt} \int_{-\infty}^\infty dx\, e^{-ikx} \Phi_i(t - |x|/B_a). \qquad (4.34)$$

We can perform a variable transformation

$$\tau = t - |x|/B_a, \qquad (4.35)$$

which leads for $\mathcal{L}$ to the integration boundaries

$$0 \to -|x|/B_a, \quad \infty \to \infty.$$

Since the reconnection electric field is zero for $t < 0$, we can introduce a condition of causuality

$$\Phi(x) \equiv 0 \quad \forall\, x < 0. \qquad (4.36)$$

Now the lower integration boundary can be replace by

$$-|x|/B_a \to 0,$$

leading to the source term

$$Q^{(1)}(p,k) = \int_{-\infty}^\infty dx\, e^{-ikx} \int_0^\infty e^{-p(\tau+|x|/B_a)} \Phi_i(\tau)d\tau = \int_{-\infty}^\infty dx\, e^{-ikx - px/B_a} \Phi(p). \qquad (4.37)$$

Integration of this equation leads to

$$\begin{aligned}
Q^{(1)}(p,k) &= \left(\frac{1}{-ik + p/B_a} e^{-ikx + px/B_a} \Big|_{-\infty}^0 + \frac{1}{-ik - p/B_a} e^{-ikx - px/B_a} \Big|_0^\infty \right) \Phi(p) = \\
&= \left(\frac{1}{-ik + p/B_a} - \frac{1}{-ik - p/B_a} \right) \Phi(p) = \\
&= \frac{2pB_a}{p^2 + B_a^2 k^2} \Phi(p) = \frac{2pB_a}{B_x(p^2 + B_a^2 k^2)} F(p), \qquad (4.38)
\end{aligned}$$

where F is the reconnected flux.

For the other three terms on the right hand side of Equation (4.33) we achieve

$$Q^{(2)}(p,k) = -\frac{2pB_b}{B_x(p^2 + k^2 B_b^2)} F(p), \qquad (4.39)$$

$$Q^{(3)}(p,k) = -\frac{2p}{p^2 + k^2 B_b^2} F(p), \qquad (4.40)$$

$$Q^{(4)}(p,k) = \frac{2p}{p^2 + k^2 B_a^2} F(p). \tag{4.41}$$

Therefore, we can write the whole source term (4.33) in $\mathcal{L}$-$\mathcal{F}$ space as

$$\begin{aligned}
Q(p,k) &= Q^{(1)}(p,k) + Q^{(2)}(p,k) + Q^{(3)}(p,k) + Q^{(4)}(p,k) = \\
&= \frac{2p}{B_x} \left(\frac{B_a - B_x}{p^2 + k^2 B_a^2} + \frac{B_b - B_x}{p^2 + k^2 B_b^2} \right) F(p) = \\
&= \frac{2 p\, v_x}{B_x} \left(\frac{1}{p^2 + k^2 B_a^2} - \frac{1}{p^2 + k^2 B_b^2} \right) F(p),
\end{aligned} \tag{4.42}$$

where we used $v_x = (B_a - B_x)/2$. Additionally, we can evaluate

$$\epsilon_+ + \epsilon_- = 2\left(p^2 + c^2 k^2\right), \tag{4.43}$$

with $c^2 = (B_a^2 + B_b^2)/2$ as the velocity of surface waves. Inserting Equations (4.42) and (4.43) into Equation (4.29) gives the displacement vector in the upper half plane as

$$\xi_{z+}(k,z,p) = \frac{p\, v_x (p^2 + k^2 B_b^2)}{B_x (p^2 + c^2 k^2)} \left(\frac{1}{p^2 + k^2 B_a^2} - \frac{1}{p^2 + k^2 B_b^2} \right) F(p)\, e^{-|k|z}, \tag{4.44}$$

and in the lower half plane as

$$\xi_{z-}(k,z,p) = -\frac{p\, v_x (p^2 + k^2 B_a^2)}{B_x (p^2 + c^2 k^2)} \left(\frac{1}{p^2 + k^2 B_a^2} - \frac{1}{p^2 + k^2 B_b^2} \right) F(p)\, e^{|k|z}. \tag{4.45}$$

4.4 Displacement vector in real space

Since we are interested in physical solutions, it is now necessary to transform the expressions for the displacement vector back into time–coordinate space. For this purpose we use the Cagniard–deHoop method, which allows to perform the inverse Fourier transformation analytically. After that, the inverse Laplace transformation can be rewritten by using the Cauchy theorem and the causality of the reconnection electric field into a convolution integral. Thus, we apply an inverse Fourier transformation $\mathcal{F}^{-1}$

$$f(x,z,t) = \frac{1}{2\pi} \int_{-\infty}^{\infty} \tilde{f}(k,z,t)\, e^{ikx} dk, \tag{4.46}$$

to Equation (4.44) giving

$$\xi_{z+}(x,z,p) = \frac{v_x}{2\pi B_x} \int_{-\infty}^{\infty} \frac{p^2 + k^2 B_b^2}{p^2 + c^2 k^2} \left(\frac{1}{p^2 + k^2 B_a^2} - \frac{1}{p^2 + k^2 B_b^2} \right) \times \\ \times p\, F(p)\, e^{-|k|z} e^{ikx} dk. \tag{4.47}$$

and perform a variable transformation $k = s\,p$ leading to

$$\xi_{z+}(x,z,p) = \frac{v_x}{2\pi B_x} \int_{-\infty}^{\infty} \frac{1+s^2 B_b^2}{1+c^2 s^2} \left(\frac{1}{1+s^2 B_a^2} - \frac{1}{1+s^2 B_b^2} \right) \times$$
$$\times F(p)\, e^{-|p\,s|z} e^{i\,s\,p\,x} ds\,. \quad (4.48)$$

If we perform a variable transformation $s \to -s$, the argument of the integral in Equation (4.48) becomes the complex conjugate, and we can replace the integration

$$\int_{-\infty}^{\infty} \to 2\,\Re \int_{0}^{\infty}.$$

This leads to

$$\xi_{z+}(x,z,p) = \frac{v_x}{\pi B_x} \Re \int_{0}^{\infty} \frac{1+s^2 B_b^2}{1+c^2 s^2} \left(\frac{1}{1+s^2 B_a^2} - \frac{1}{1+s^2 B_b^2} \right) F(p)\, e^{-p\,\tau(s)} ds\,, \quad (4.49)$$

with $\tau(s) = s\,z - i\,s\,x$. The argument of the integral in Equation (4.49) has the same form like the shift theorem for the Laplace transformation

$$\mathcal{L}(\Phi(t-a)) = e^{-p\,a} \hat{\Phi}(p) \quad \text{for } a \in \mathbb{R}. \quad (4.50)$$

In our case, it is obvious that $\tau(s) \in \mathbb{C}$, but we can use the Cauchy theorem to transform $\tau(s)$ to a curve, where $\Im(\tau(s)) = 0$. Cauchy theorem proposes that if $\mathcal{C}$ is a closed path lying within a region, where $f(z)$ is analytic, meaning that there are no poles inside this region, then

$$\oint_{\mathcal{C}} f(z) dz = 0\,. \quad (4.51)$$

Now we can construct a closed path, which is shown in Figure 4.2, where $\mathcal{C}_1$ represents the integration from 0 to infinity. Now we can define a curve, where $\Im(\tau(s)) \equiv 0$. Substituting $s = s_r + i\,s_i$ gives

$$\Im\left(-i\,(s_r + i\,s_i)\,x + (s_r + i\,s_i)\,z\right) = 0\,, \quad (4.52)$$

which leads to

$$\frac{s_i}{s_r} = \frac{x}{z}\,. \quad (4.53)$$

This integration path is represented by $\mathcal{C}_2$ in Figure 4.2. It can be shown that $\Re(\tau(s)) > 0$ along $\mathcal{C}_3$, and therefore this integral is decreasing exponentially and can be neglected. Due to Equation (4.51), we can replace the integration along $\mathcal{C}_1$ by an integration along $\mathcal{C}_2$, where $\Im(\tau(s)) = 0$.

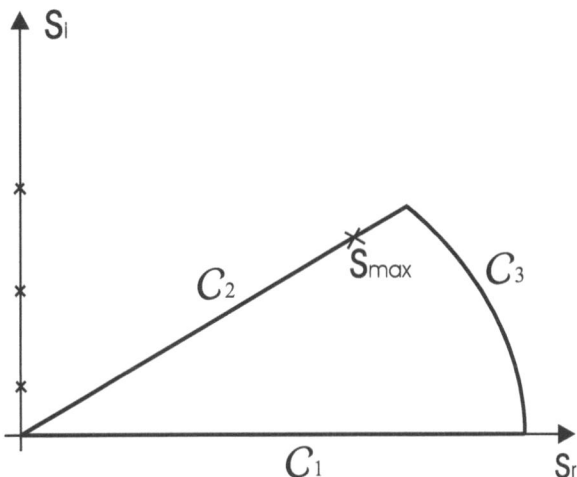

Figure 4.2: *Closed path for the evaluation of the Cauchy theorem. The crosses indicate that the poles are located along the imaginary axis.*

After application of an inverse Laplace transformation, Equation (4.49) can be written as

$$\xi_{z|}(x,z,t) = \frac{v_x}{\pi B_x} \Re \int_{C_2} \frac{1+s^2 B_b^2}{1+c^2 s^2} \left(\frac{1}{1+s^2 B_a^2} - \frac{1}{1+s^2 B_b^2} \right) \times$$
$$\times \frac{1}{2\pi i} \int_{\sigma-i\infty}^{\sigma+i\infty} F(p) \, e^{-p\tau(s)} e^{pt} dp \, ds \,. \quad (4.54)$$

Since $\tau(s) \in \mathbb{R}$, we can use the shift theorem for Laplace transformations (4.50) and get

$$\xi_{z+}(x,z,t) = \frac{v_x}{\pi B_x} \Re \int_{C_2} \frac{1+s^2 B_b^2}{1+c^2 s^2} \left(\frac{1}{1+s^2 B_a^2} - \frac{1}{1+s^2 B_b^2} \right) F(t-\tau(s)) \, ds \,. \quad (4.55)$$

Due to the condition of causality (4.36), $F(t - \tau(s)) \equiv 0$ if $t - \tau(s) \leq 0$. Therefore, we can define an upper limit s_{max} (Figure 4.2) for the integration along C_2 when $t = \tau(s)$. After performing a variable transformation $s \to \tau$, we get the displacement vector in real space as the convolution integral

$$\xi_{z+}(x,z,t) = \frac{v_x}{\pi B_x} \Re \int_0^t f_+(\tau) F(t-\tau) \frac{1}{q_+} d\tau \,, \quad (4.56)$$

with
$$f_+(\tau) = \frac{q_+^2 + B_b^2\tau^2}{q_+^2 + c^2\tau^2}\left(\frac{q_+^2}{q_+^2 + B_a^2\tau^2} - \frac{q_+^2}{q_+^2 + B_b^2\tau^2}\right), \quad (4.57)$$
and
$$q_+ = z - ix. \quad (4.58)$$

The same arguments give the displacement vector in the lower half plane as
$$\xi_{z-}(x,z,t) = -\frac{v_x}{\pi B_x}\Re\int_0^t f_-(\kappa)F(t-\kappa)\frac{1}{q_-}d\kappa, \quad (4.59)$$
with
$$f_-(\kappa) = \frac{q_-^2 + B_a^2\kappa^2}{q_-^2 + c^2\kappa^2}\left(\frac{q_-^2}{q_-^2 + B_a^2\kappa^2} - \frac{q_-^2}{q_-^2 + B_b^2\kappa^2}\right), \quad (4.60)$$
and
$$q_- = -z - ix. \quad (4.61)$$

4.5 Evaluation of the MHD quantities

Knowing the analytical expression for the displacement vector, it is possible to derive all MHD quantities from it. To calculate the z–component of the magnetic field, we use that
$$B_{z+} = B_a\frac{\partial}{\partial x}\xi_{z+}. \quad (4.62)$$

It is convenient to perform the spatial derivation in the $\mathcal{L}$-$\mathcal{F}$ space, so that we can insert Equation (4.48) into Equation (4.62), yielding
$$B_{z+} = \frac{v_x B_a}{2\pi B_x}\int_{-\infty}^{\infty}\frac{1+s^2 B_b^2}{1+c^2 s^2}\times$$
$$\times\left(\frac{1}{1+s^2 B_a^2} - \frac{1}{1+s^2 B_b^2}\right)F(p)\,i\,s\,p\,e^{-|p s||z|}e^{i s p x}ds. \quad (4.63)$$

In Laplace space the reconnection electric field can be written as $E(p) = p F(p)$, which gives the z–component of the magnetic field as
$$B_{z+}(x,z,t) = \frac{v_x B_a}{\pi B_x}\Re\int_0^t f_+(\tau)E(t-\tau)\frac{i\tau}{q_+^2}d\tau. \quad (4.64)$$

The x–component of the magnetic field can be determined by using the definition of the displacement vector (4.17), which gives for the x–component
$$B_{x+} = B_a\frac{\partial}{\partial x}\xi_x. \quad (4.65)$$

If we use that the divergence of the displacement vector is zero, we can write Equation (4.65) as

$$B_{x+} = -B_a \frac{\partial}{\partial z}\xi_z. \tag{4.66}$$

If we solve this equation in $\mathcal{L}$-$\mathcal{F}$ space, we get the x-component of the magnetic as

$$B_{x+}(x,z,t) = \frac{v_x B_a}{\pi B_x} \Re \int_0^t f_+(\tau) E(t-\tau) \frac{\tau}{q_+^2} d\tau. \tag{4.67}$$

The calculated profiles (upper panel of Figure 4.3) are equivalent to profiles calculated by Fourier method.

For the z-component in the lower half plane we use

$$B_{z-} = B_b \frac{\partial}{\partial x}\xi_{z-}. \tag{4.68}$$

If we repeat the procedure described above with ξ_{z-}, the z-component of the magnetic field can be written as

$$B_{z-}(x,z,t) = -\frac{v_x B_b}{\pi B_x} \Re \int_0^t f_-(\tau) F(t-\tau) \frac{i\tau}{q_-^2} d\tau. \tag{4.69}$$

For the x-component this procedure can be repeated analogously.

The z-component of the velocity in the upper half plane is given as

$$v_{z+} = \frac{\partial}{\partial t}\xi_{z+}. \tag{4.70}$$

We can perform the temporal derivation in $\mathcal{L}$ space and get the z-component of the velocity as

$$v_{z+}(x,z,t) = \frac{v_x}{\pi B_x} \Re \int_0^t f_+(\tau) E(t-\tau) \frac{1}{q_+} d\tau. \tag{4.71}$$

The x-component of the velocity is given as

$$v_{x+}(x,z,t) = \frac{\partial}{\partial t}\xi_{x+}. \tag{4.72}$$

The x-component of the displacement vector can be evaluated by using $\nabla \cdot \xi = 0$, which gives in $\mathcal{L}$-$\mathcal{F}$ space

$$ik\xi_x + \frac{\partial}{\partial z}\xi_z = 0. \tag{4.73}$$

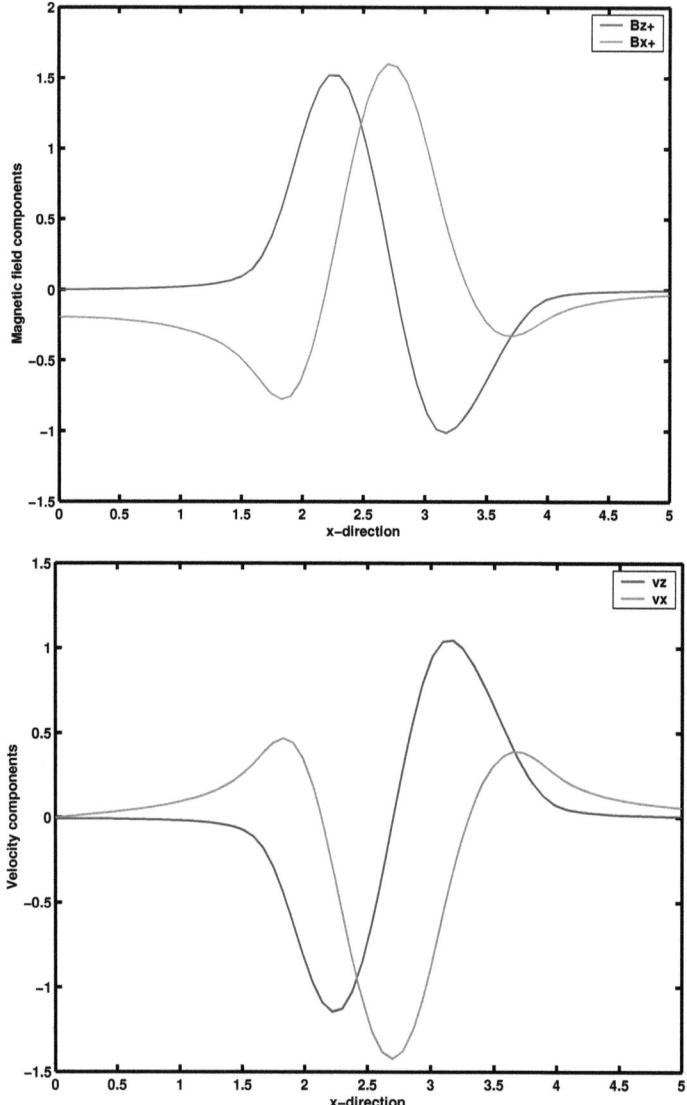

Figure 4.3: *Behavior of the $x-$ and $z-$components of the magnetic field (upper panel) and the velocity (lower panel) in the upper half plane calculated via Cagniard-deHoop method for $B_a = 2$, $B_b = -1$, $z = 0.1$, and $t = 2$.*

Since we know ξ_z, we get the x-component of the displacement vector as

$$\xi_{x+}(x,z,t) = \frac{v_x}{\pi B_x}\Re \int_0^t f_+(\tau) F(t-\tau) \frac{1}{i\, q_+} d\tau, \quad (4.74)$$

and the x-component follows as

$$v_{x+}(x,z,t) = \frac{v_x}{\pi B_x}\Re \int_0^t f_+(\tau) E(t-\tau) \frac{1}{i\, q_+} d\tau. \quad (4.75)$$

This is shown in lower panel of Figure 4.3.
To get the total pressure, we can integrate Equation (4.19) giving

$$P_+ = \frac{\epsilon_+ \epsilon_-}{\epsilon_+ + \epsilon_-} \frac{1}{|k|} Q(k,p) e^{-|k|z}. \quad (4.76)$$

Now we can substitute the source term (4.42) and apply the Cagniard-deHoop method, getting the total pressure in the upper half plane as

$$P_+(x,z,t) = \frac{v_x}{\pi B_x}\Re \int_0^t f_+(\tau) E(t-\tau) \frac{q^2 + B_a^2 \tau^2}{q^2 \tau} d\tau. \quad (4.77)$$

4.6 Magnetic field along trajectories

The previous plots were made for a fixed point in time, but in reality the spacecraft is nearly fixed, since the velocity of the spacecraft is only some km/s, while the disturbances are propagating with velocities of some 100 km/s. Therefore, it is more realistic to consider the case with a spacecraft at a fixed position and with changing time (Figure 4.4). All analysis of FTEs and NFTEs starting from the pioneer work of Russell and Elphic (1978) were done for magnetic field trajectories. As it can be seen from Fig. 4.4, all features expected for FTEs in the incompressible case can be found from the presented model. The used electric field is of the form

$$E(t) = \frac{b^2 e^2}{20} t^2 e^{-bt}, \quad (4.78)$$

with $b = 4$. The characteristic asymmetric bipolar variation expected for the B_z-component is clearly visible, and also the deflection of the x-component of the magnetic field. Additionally, the velocity behaves as proposed for NFTEs, with a upward flow of plasma in the beginning, followed by a strong flow directed the downward to the plasma sheet. Moreover, the maximum of the B_x-component is correlated with a change in sign of the other components, which is a characteristic feature for NFTEs (Sergeev et al., 1992,

Figure 4.4: *Behavior of the x–component (dashed line) and the z–component (solid line) of the magnetic field perturbations (upper panel), and the z–component of the plasma velocity (lower panel), which are used to define FTEs/NFTEs. The bipolar variation of B_z and the anticorrelated variation of v_z is evident. The vertical dashed–dotted line indicates the time point, where B_x has a maximum and the other two quantities are changing their sign (from Penz et al., 2006a).*

2005; Penz et al., 2006a), which is indicated by the vertical dashed–dotted line.
The bipolar variation of the normal component B_z can be explained by considering a bulge passing by the satellite (for a magnetopause context see Russell and Elphic (1978)). First, one can observe a change of the normal component reaching a maximum at the steepest slope of the bulge followed by a return to zero when the bulge reaches its maximum displacement. After that the bulge is propagating away from the satellite position, leading to a perturbation in the opposite direction and a return to zero afterwards. Since the bulge is associated with reconnection in our model, the downward motion is also caused by reconnection, leading to a slightly asymmetric bipolar variation. The motion of the plasma can be interpreted in the following way: the leading edge of the shock front is causing a shift of the plasma away from the current sheet, giving the upward plasma flow. After the bulge passed by,

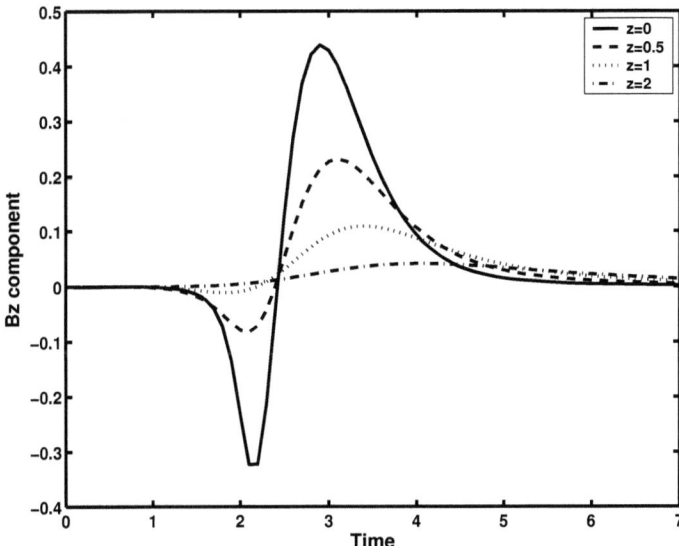

Figure 4.5: *Time series of the z–component of the magnetic field perturbations calculated at $z = 0$ (solid line), $z = 0.5$ (dashed line), $z = 1$ (dotted line), and $z = 2$ (dashed dotted line) (from Semenov et al., 2005a).*

a convective flow directed to the plasma sheet can be observed. In Figure 4.5, the z–component of the magnetic field perturbations for different z–distances of the observer is shown. For large normalized distances of $z > 1$ (to achieve the dimensional distance, the values must be multiplied by $v_A T^{(0)}$), the features are becoming small, which limits the method to distances smaller than about $z = 2$.

5 Reconstruction methods and the inverse problem

In the previous section, we derived analytical expressions for the MHD quantities in the form of convolution integrals of the reconnection electric field with an integration kernel. This section is devoted to the reconstruction of the reconnection electric field by applying an inverse model. A method to solve inverse problems where convolution integrals are involved is Tikhonov regularization. We use this method to solve the inverse problem and we reconstruct the reconnection electric field for different magnetic field perturbations generated via the direct method. This enables us to find the limitations of the inverse model.

5.1 Reconstruction methods for reconnection features

After the observation of FTE signatures, some attempts were made to reconstruct different features of the reconnection process involved in the generation of FTEs. Southwood (1985) predicted that FTE signatures would be observed by a satellite regardless of whether or not it actually penetrates the FTE. He studied the hydromagnetic aspects of the situation involved, and noted that the forces acting on the connected flux tube differ from those on the surrounding media. Consequently, the tube must move with a different velocity compared with the ambient plasma. The surrounding plasma must then move out of the way of the tube, which leads to exterior flow perturbations and hence, via the frozen–in field condition, to field perturbations that are also external to the connected tube. Farrugia et al. (1987) verified Southwood's suggestion and reproduced the magnetic field signatures outside the flux tube by considering the flow of an inviscid, incompressible plasma over a semi-circular cylinder. They showed that FTE–like signatures could be generated without the satellite is penetrating the obstacle, including all expected features like the draping of all three components of the magnetic field and accompanying changes in the field strength. Also the perturbations in the plasma velocity are related to those in the magnetic field and there are concomitant pressure disturbances as well. A detailed interpretation of one magnetosheath FTE in terms of the Farrugia et al. (1987) model was offered by Papamastorakis et al. (1989). These authors observed that the magnetic field perturbations derived from Farrugia et al. (1987) produce a cardioid–shaped magnetic hodogram in the plane perpendicular to the flux tube axis (the x–z–plane in the notation used in this work) with no magnetic field perturbations along that axis (the y–axis in the notation used in this work). It follows from this latter property that the axis orientation can be found by Minimum Variance Analysis (MVA) of the measured magnetic field (which would ideally display zero variance in its component along the y–axis). The actual hodogram found by Papamastorakis et al. (1989) deviated significantly from a cardioid, which could be interpreted to mean that the actual cross section of the tube was not semicircular as proposed by Farrugia et al. (1987).

Based on this discrepancy, Walthour et al. (1993) developed a technique based on a linear theory for isentropic field–aligned MHD flow over gently sloping two–dimensional obstacles using integral transforms. Using only magnetic field measurements, the analysis technique can provide information about the orientation and actual cross–sectional shape of the event (Figure 5.1a), as well as information about the spacecraft trajectory relative to the bulge. This method is based on an assumption, which was also an essential part of the model by Farrugia et al. (1987), namely that a so–called deHoffmann–Teller (HT) frame exists in which the flow is field–aligned. A principal implication of the existence of an HT frame is that the magnetic field is time–independent when viewed in that frame. In other words, the observed magnetic structures are coherent on time scales shorter or equal to the duration of the event. The analysis of two sample events, one recorded by AMPTE/IRM in the magnetosheath and the other by AMPTE/CCE in the magnetosphere, indicates that the bulges on the magnetopause surface causing the magnetic field and flow perturbations did not have a semicircular cross section. Instead they had an elongated shape, with the dimension tangential to the magnetopause being substantially larger than that normal to it. Walthour et al. (1994) confirmed this finding by analyzing an FTE recorded by ISEE 1 and 2 in the northern hemisphere near local noon. Since the analysis is confined to perturbations outside the obstacle, the method is referred to as a remote sensing method.

Another approach to this topic was used by several authors (Hau and Sonnerup, 1999; Hu and Sonnerup, 2001; 2003) who developed a method based on the Grad–Shafranov (GS) equation to reconstruct two-dimensional space plasma structures in magnetohydrostatic equilibrium (Figure 5.1b). The technique involves solving the plane ($\partial/\partial z \approx 0$) GS equation in Cartesian coordinates (in SI units)

$$\frac{\partial^2 A}{\partial x^2} + \frac{\partial^2 A}{\partial y^2} = -\mu_0 \frac{dP_t}{dA} = -\mu_0 j_z(A),$$

where the magnetic vector potential, $A(x,y)\,\mathbf{z}$, is defined suc that the magnetic field $\mathbf{B} = (\partial A/\partial y, -\partial A/\partial x, B_z(A))$. The transverse pressure, $P_t(A) =$

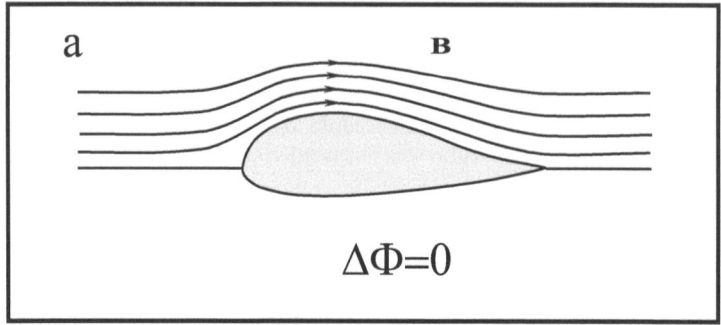

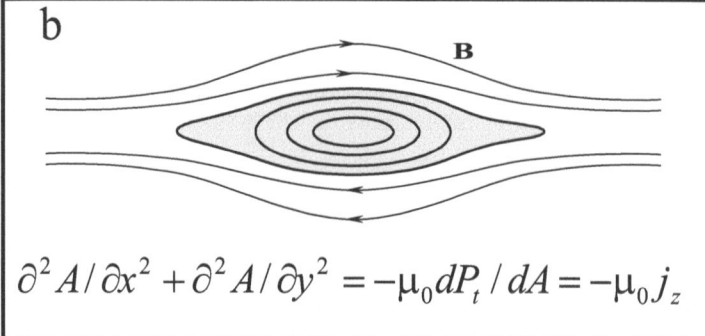

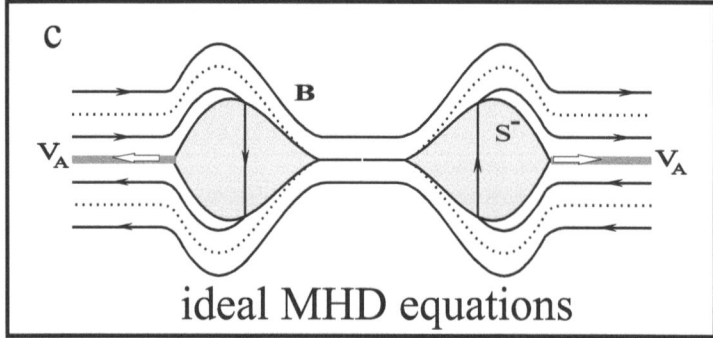

Figure 5.1: *Different approaches to describe and analyze FTEs: a) is the potential flow method used by Walthour et al. (1993, 1994), b) is the approach using the Grad–Shafranov equation to calculated structures in magnetohydrostatic equilibrium (Sonnerup et al., 2004), and c) is the model of transient Petschek-type reconnection (Biernat et al., 1987; Semenov et al., 1992). It can be seen that only the last approach is directly associated with the reconnection of magnetic field lines and therefore allows to reconstruct the reconnection rate.*

$p(A) + B_z^2(A)/2\mu_0$, and therefore the axial current density $j_z(A)$ are functions of A alone. The magnetic variance analysis and deHoffmann–Teller analysis of measured magnetic field vectors $\mathbf{B}$ and plasma flow velocities $\mathbf{v}$ are performed. A normal vector $\mathbf{n}$ to the magnetopause is determined as the minimum variance direction. A constant HT frame velocity $\mathbf{v}_{HT}$ is determined from standard HT analysis. The magnetic structure is then assumed to convect past the spacecraft with velocity $\mathbf{v}_{HT}$. An optimal z-axis is found in the plane perpendicular to the normal vector by trial and error, rotating the trial z-axis stepwise away from its initial direction, such that the requirements, $P_t = P_t(A)$, and/or $p = p(A)$ and $B_z = B_z(A)$ are optimally satisfied for data points located on the same field line. Viewed in a co–moving frame, the spacecraft moves across the magnetic structures along a straight line with the velocity $-\mathbf{v}_{HT}$. The magnetic potential A, at points along the projected spacecraft trajectory, i.e., the x-axis, is calculated by integration of the measured B_y component as

$$A(x,0) = \int_0^x \frac{\partial A}{\partial \xi} d\xi = \int_0^x -B_y(\xi,0)\, d\xi.$$

The space increment along the x-axis is obtained from the time increment via the constant deHoffmann–Teller velocity $d\xi = -\mathbf{v}_{HT} \cdot \mathbf{x} dt$. The magnetic field components are spacecraft measurements projected from an observational frame, such as the Cartesian GSE coordinate system onto the (x, y, z) system. Since they and the pressure $p(x, 0)$ are known on te x-axis owing to the invariance in the z direction, a functional fit of $P_t(x, 0)$ versus $A(x, 0)$ is used to approximate $P_t(A)$ on the right–hand side of the GS equation. Then the integration proceeds explicitly in y direction starting at $y = 0$, utilizing $B_x(x, 0) = \partial A/\partial y|_{y=0}$ and $A(x, 0)$ as initial values. Thus the magnetic potential value $A(x, y)$ is obtained in a rectangular domain surrounding the x-axis. The contour plot of $A(x, y)$, called a field map, represents the transverse magnetic field lines. The field component B_z as well as the plasma pressure p is calculated by evaluating functions $B_z(A)$ and $p(A)$, obtained by fitting to the spacecraft data at points along the trajectory. Using this method, the authors are able to visualize two–dimensional, time–independent magnetic structures by using single satellite measurements. Hu and Sonnerup (2003) applied this model to two magnetopause crossings by the spacecraft AMPTE/IRM, and reconstructed magnetic field structures. Sonnerup et al. (2004) used this method to give a rough estimation of an average reconnection rate. Hasegawa et al. (2004) applied this method to Cluster measurements and compared the magnetic field maps produced from one spacecraft with the field vectors that the other spacecrafts actually observed. For an optimally selected invariant axis, the correlation between the field components predicted from the reconstructed map and the corresponding measured components reaches more than 0.97. This indicates that the reconstruction technique predicts conditions at the other spacecraft locations quite well. Hasegawa et al. (2005) improved the method to a multi–spacecraft technique that produces a single field map by ingesting data from all four Cluster spacecraft into the calculation. The plasma pressure required for the technique is measured in high time resolution by only two of the spacecraft, but with the help of spacecraft potential measurements available at all four spacecraft, the pressure can be estimated also at the other spacecraft. Consequently, four independent field maps can be reconstructed and then merged into a single map. The resulting map appears more accurate in the sense that the agreement between magnetic field variations predicted from the map to occur at each of the four spacecraft and those actually measured is significantly better. Such a composite map does not satisfy the GS equation anymore, but is optimal under the constraints that the structures are two–dimensional and time–independent. However, the application of this method to the propagation of a reconnected flux tube is limited since such a propagation can rather be described by the inertia force of the plasma and the Maxwellian tension, which is essentially a time–dependent process, and therefore this process can be hardly understood in a magnetohydrostatic approach.

Recently, different models based on multi–spacecraft measurements were developed to determine the position of the reconnection site with respect to the satellite. Wild et al. (2005) used data from Cluster and Geotail positioned at the high– and the low–latitude dayside magnetopause. In order to study the observed FTE features, they employed a model of open flux tube motion to investigate the motion of newly reconnected flux tubes over the dayside magnetopause. This model considers the draping and strength of the magnetosheath magnetic field, magnetosheath flow velocity, and density over the surface of a simple paraboloid magnetopause. Wild et al. (2005) are able to monitor the subsequent motion of the reconnected flux tubes away from a user–defined reconnection site. By doing this, they are able to compare the expected flux tube motion with the observed FTE signatures without constraining the location of the reconnection site, using an assumed threshold to the reconnection process. Furthermore, rather than imposing a particular X–line orientation, they simply trace the flux tube motion from several point-like reconnection sites in order to mimic the effects of a spatially extended X–line. Wild et al. (2005) applied this model to a series of FTEs observed on February 17th, 2003, by the Cluster and Geotail satellites in the vicinity of the high– and low–latitude dayside magnetopause, giving a reconnection site located at near–equatorial latitudes. In another approach, Fuselier et al. (2005) determine the reconnection inflow velocity into the magnetosphere, and use Cluster and IMAGE data to estimate the approximate reconnection site.

In order to describe the temporal evolution of FTEs, time–dependent Petschek–type models of reconnection were developed (e.g., Biernat et al., 1987; Semenov et al., 1992, 2004a, 2004b; Heyn and Semenov, 1996). Figure 5.1c shows the switch–off stage in the time-dependent model of magnetic reconnection. Here, reconnection is initiated by the appearance of an electric field in a localized region of the current sheet separating oppositely directed magnetic fields. Due to this process, reconnection–associated disturbances are propagating along the current sheet in the form of pairs of shocks moving in opposite direction away from the reconnection site. These shocks bound a region of heated and accelerated plasma, the so–called outflow region (the shaded areas in Fig. 5.1c). After the electric field has dropped to zero, the shocks detach from the reconnection site and continue to propagate along the current sheet, which can be seen in Fig. 5.1c. It should be noted that even the reconnection process ceased, energy conversion still takes place, leading to a continued growth of the outflow region. Lawrence et al. (2000) analyzed a series of FTE-like events generated by a time–dependent model of reconnection by using the remote sensing method developed by Walthour et al. (1993, 1994). They used two– and three–dimensional versions of a time–dependent Petschek–type model of reconnection, where the FTE features are interpreted as perturbations in the ambient magnetic field and plasma flow produced by the growth and propagation of an outflow region of reconnected plasma. Penz (2002) used a representation in the form of Poisson integrals and alternatively one based on Fourier transformations to describe the perturbations caused by a reconnection electric field in the ambient plasma environment along a profile. Solving an inverse problem using Tikhonov regularization, this method allows to reconstruct the reconnection electric field from mathematically generated FTE signatures. However, satellite measurements are not profiles but trajectories. Therefore, Semenov et al. (2005b) used an alternative approach based on the Cagniard–deHoop method and reconstructed electric fields with different shapes. Semenov et al. (2005a) and Penz et al. (2004, 2005, 2006a, 2006b) applied this method to several NFTEs measured by the Cluster satellites.

5.2 The inverse problem in an incompressible plasma

Inverse problems are well–known in many fields of science and technology, for example in astrophysics, plasma diagnostics or tomography (Tikhonov and Goncharsky, 1987). In principle, an inverse problem can be seen as a model of a phenomenon characterized by ζ, which belongs to a certain space of models Z. Furthermore, let u be the observed indirect attributes of the phenomenon with $u \in U$. An operator A relates the two values by

$$A\zeta = u.$$

As a rule, the ζ attributes cannot be directly observed.
The main job when solving an inverse problem is to find out, whether the chosen model is compatible with the experimental data. The mathematical difficulty of solving such problems is that the inverse operator A^{-1} defined throughout its domain $AZ \subset U$ is not continuous. Hence there is the conventional division into the classes of well–posed and ill–posed problems. A well–posed problem, as defined by Hadamard (1932), must meet the requirements that:

1. equation $A\zeta = u$ is solvable for the entire space U,

2. the solution is unique,

3. the solution is stable, meaning that small perturbations of u result in small perturbations of the solution.

For ill–posed problems, requirements 1 and 3 are not fulfilled. These kind of problems can be solved by using the theory of regularization (Tikhonov and Arsenin, 1977), which we apply in the following to our problem.
In nature, the outflow regions are moving with velocities of some hundred km/s, while the satellite's velocity is only some km/s. Therefore, we can consider the satellite as fixed, meaning that $x = const$ and $z = const$ in this case. Now the magnetic field is only a function of time $B_z(x, z, t) \equiv B_z(t)$. Our starting point is the convolution integral for the z–component of the magnetic field (4.64) for a fixed position in space. In Laplace space, the convolution integral can be written as

$$B_z(p) = K(p)E(p). \qquad (5.1)$$

From Equation (5.1) it looks quite simple to reconstruct the reconnection electric field as

$$E(p) = \frac{B_z(p)}{K(p)}. \qquad (5.2)$$

If we visualize both functions $K(p)$ and $B_z(p)$ (Figure 5.2), we see that for large absolute values of z we divide very small numbers. This leads to large oscillations in the result, which is a characteristic feature for ill–posed inverse

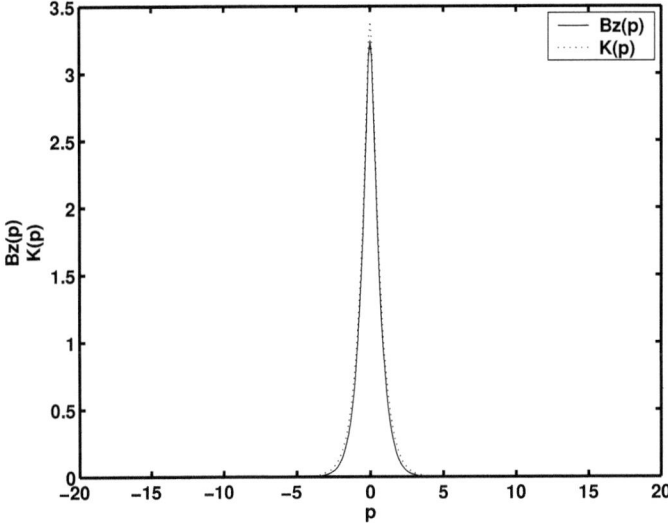

Figure 5.2: *Behavior of $K(p)$ and $B_z(p)$ in Laplace space.*

problems. To avoid this oscillating function (Figure 5.3), we introduce again a regularization operator $M(p)$ in Equation (5.2) giving

$$E(p) = \frac{B_z(p)}{K(p) + M(p)} . \qquad (5.3)$$

The regularization operator $M(p)$ is defined as

$$M(p) = \begin{cases} 0 & |p| < R_{max} \\ \infty & |p| > R_{max} \end{cases} . \qquad (5.4)$$

This operator does not influence the electric field for small values of p, where the two functions are not small. But when the functions reach small values ($p > R_{max}$), the denominator goes to infinity, so that the reconnection electric field is zero in Laplace space and large oscillations are suppressed. The value of R_{max} is found from internal parameters of the numerical Laplace transformation.

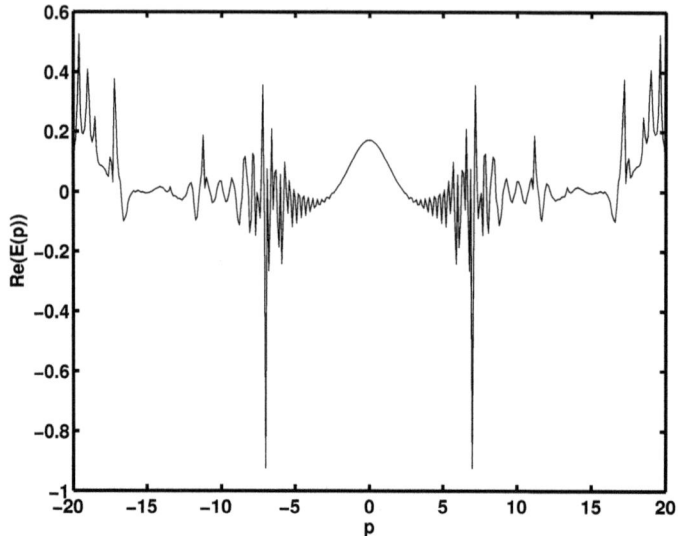

Figure 5.3: *Real part of the reconnection electric field without regularization. In this case, $R_{max} \approx 4$.*

5.3 Reconstruction of an exponential reconnection electric field

At first, we use an initial electric field of the form

$$E(t) = \frac{b^2 e^2}{20} t^2 e^{-bt}, \qquad (5.5)$$

with $b = 4$. We can use the magnetic field trajectories obtained via the Cagniard–deHoop method to reconstruct the electric field. If we use a magnetic field configuration with $B_a = 2$ and $B_b = -1$ and reconstruct the electric field from a satellite position at $x = 5$ and $z = 2$ the result is very satisfying (Figure 5.4). The coincidence between the initial and the reconstructed electric field is good and only small oscillations occur. The value of R_{max} is chosen to be 7.5 in this case. For a larger distance of the satellite above the reconnection site of $z = 5$ the agreement is still qualitatively good with $R_{max} = 5$. Therefore, this remote sensing method works good for distances up to more than 50 times the height of the outflow region. This shows that the assumption of weak reconnection is justified since it leads to a thin outflow region, but also in this case it works for quite large absolute

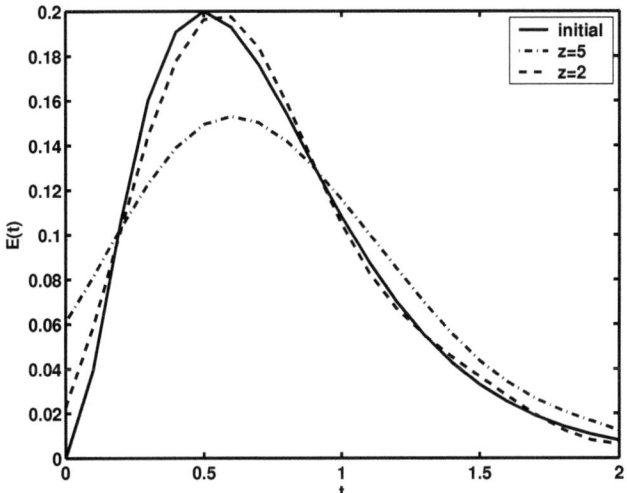

Figure 5.4: *Reconstructed and initial electric field for a position $x = 5$ and $z = 2$ (dotted line) and $z = 5$ (dashed-dotted line) for a magnetic field configuration $B_a = 2$ and $B_b = -1$.*

distances. For larger distances above the reconnection site the agreement is getting worse. The value of x does not influence the agreement significantly. Now we can vary the magnetic field configuration. The situation for $B_a = 5$ and $B_b = -1$ is shown in Figure 5.5. For a distance of $z = 2$ we use $R = 10$ and can reconstruct the electric field very good. For a larger distance of $z = 5$ and $R = 5$ the electric field can be restored better than for a magnetic field of $B_a = 2$. Also for a magnetic field configuration of $B_a = 5$ and $B_b = -4$, the reconstruction method works very good.

5.4 Reconstruction of sine–shaped electric fields

In this case, we use an initial electric field as

$$E(t) = \begin{cases} \sin^2(\pi t) & 0 \le t \le 1 \\ 0 & \text{else} \end{cases}. \tag{5.6}$$

The reconstructed electric field for a magnetic field configuration of $B_a = 2$ and $B_b = -1$ is shown in Figure 5.6. For a distance of $z = 2$ above the reconnection site the agreement is quite good by using $R = 10$, but for $z = 5$ the coincidence is worse than for an exponential electric field since the sine

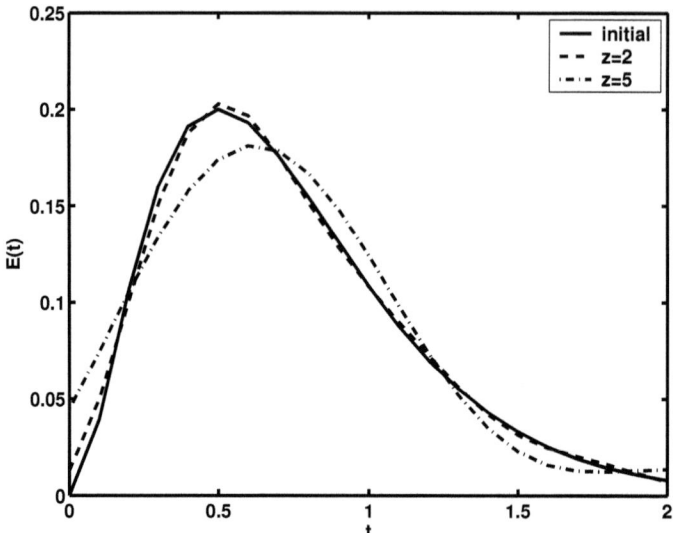

Figure 5.5: *Same as Figure 5.4, but for a magnetic field configuration $B_a = 5$ and $B_b = -1$.*

pulse is not as smooth as the exponential function. For a larger magnetic field in the upper half plane $B_a = 5$, the initial magnetic field can be reconstructed very well also for a distance of $z = 5$. Additionally, we can model the case of two reconnection pulses. Here we use a initial reconnection electric field of the form

$$E(t) = \begin{cases} \sin^2(\pi t) & 0 \leq t \leq 1 \quad \text{and} \quad 2 \leq t \leq 3 \\ 0 & \text{else} \end{cases}. \quad (5.7)$$

If we apply our reconstruction method to such an electric field, we can reconstruct the electric field for $z = 2$ very good and for $z = 5$ qualitatively good (Figure 5.7) for a magnetic field $B_a = 2$ and $B_b = -1$. The used values for R are 15 for $z = 2$ and 7.5 for $z = 5$. As in the previous cases, the magnetic field configuration of $B_a = 5$ and $B_b = -1$ enables us to reconstruct the electric field better than for $B_a = 2$. If the time separation of the pulses is larger, the method works similar than for single pulses. For large distances and small time separation it is not possible to reconstruct the right structure from the magnetic field profiles since they are not clearly separated at large distances above the reconnection site.

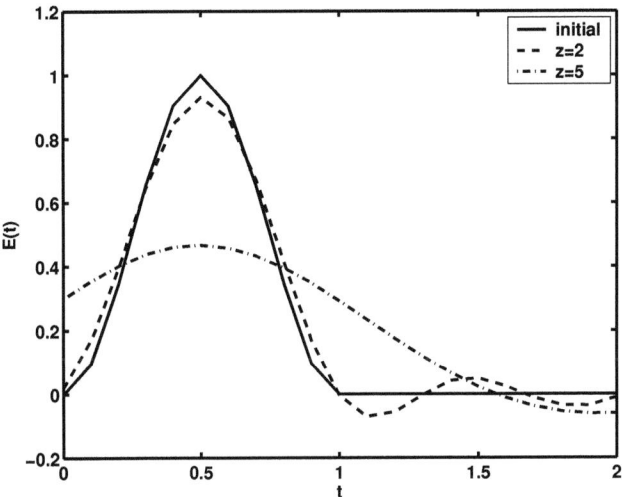

Figure 5.6: *Reconstructed and initial electric field for a sine–shaped electric field for $B_a = 2$ and $B_b = -1$ for a distance of $z = 2$ (dotted line) and $z = 5$ (dashed–dotted line).*

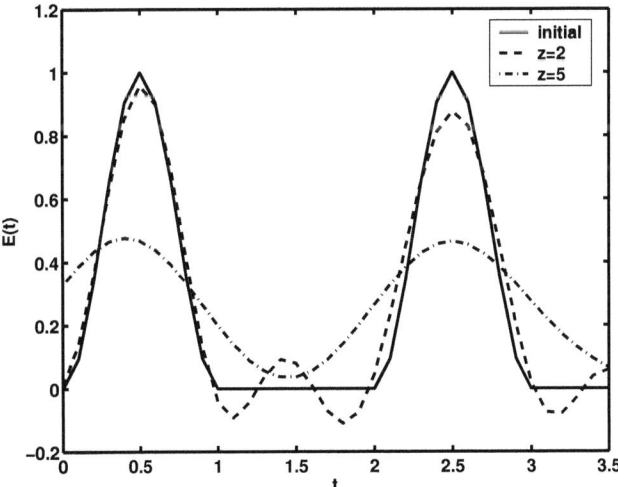

Figure 5.7: *Reconstructed and initial electric field two sine–shaped reconnection pulses for $B_a = 2$ and $B_b = -1$ for a distance of $z = 2$ (dotted line) and $z = 5$ (dashed–dotted line).*

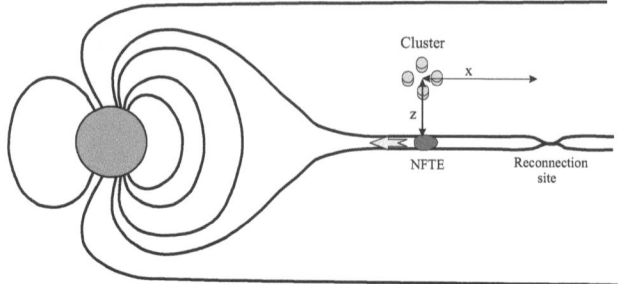

Figure 5.8: *Position of the Cluster satellites with respect to the reconnection site. The z–distance is found from the modelling done by Sergeev et al. (2005), while the x–distance is reconstructed by using a minimization routine.*

5.5 Reconstruction of the reconnection site

Since we want to apply the reconstruction method also to satellite data, an additional problem appears. Looking at Eq. 5.3, one can see that it depends on the integration kernel $K(p)$. From Eq. 4.63 it is evident that the kernel depends on the magnetic field configuration and on the location of the reconnection site with respect to the point of observation (Fig. 5.8). For the magnetic field configuration we assume that it is symmetric and antiparallel, where the values for the magnetic field components are known from satellite measurements. The z–distance between the observing satellite and the reconnection site can be found from magnetotail modelling (e.g., Kubyshkina et al., 2002; Sergeev et al., 2005). The determination of the x–distance is done by using a global minimization routine. In our time–dependent model this is possible, because the shape of the shock is changing if the x–distance from the reconnection site is increasing. We use the measured B_z component as an input data, calculate the electric field in Laplace space, which should be strictly positive, but since we do not know the exact x–distance it can appear negative somewhere. Therefore, we take the module of the electric field and recalculate $\tilde{B}_z$ out of it. This procedure can be summarized as

$$B_z(t) \Rightarrow B_z(p) \Rightarrow E(p) \Rightarrow E(t) \Rightarrow |E(t)| \Rightarrow \tilde{B}_z(t)\,.$$

Then we minimize the difference between $B_z(t)$ and $\tilde{B}_z(t)$ with a least square approach in order to find the x–distance as the value of x where the difference between the initial and the reconstructed magnetic field has a minimum. We limit the search to distances less than 35 R_e, which corresponds to the region of the near Earth neutral line (NENL), where reconnection most likely takes place. The local velocity of the disturbances is determined by using multipoint timing analysis (Harvey, 1998). We assume that this velocity is approximately the Alfvén velocity. Additionally, multipoint timing analysis gives the direction of propagation of the disturbances, which can be used to verify if they are propagating mainly in x–direction, since this is a preferable configuration for our 2–D model. Knowing all these parameters, it is possible to analyze events measured by satellites, which will be discussed in the following chapter.

6 Applications of the incompressible method to Cluster measurements in the Earth's magnetotail

6.1 The Cluster mission

The aim of the Cluster mission is to study small-scale structures of the magnetosphere and its environment in three dimensions. To achieve this, Cluster is constituted of four identical spacecraft (Rumba, Samba, Salsa, and Tango) that will flight in a tetrahedral configuration. The separation distances between the spacecraft will be varied between 600 km and 20000 km, according to the key scientific regions. Cluster is part of an international collaboration to investigate the physical connection between the Sun and Earth. Flying in a tetrahedral formation, the four spacecraft collect the most detailed data yet on small–scale changes in near–Earth space and the interaction between the charged particles of the solar wind and Earth's atmosphere. The satellites will follow highly elongated, polar orbits which take them between 19000 and 119000 kilometers from the planet. The principal parameters of the four identical satellites is shown in Table 6.1.

Each of the four spacecraft carries an identical set of 11 instruments to investigate charged particles, electrical and magnetic fields. Namely, these are the Fluxgate Magnetometer (FGM), the Electron Drift Instrument (EDI), the Active Spacecraft Potential Control experiment (ASPOC), the Spatio-Temporal Analysis of Field Fluctuation experiment (STAFF), the Electric Field and Wave experiment (EFW), the Digital Wave Processing experiment (DWP), the Waves of High frequency and Sounder for Probing of Electron density by Relaxation experiment (WHISPER), the Wide Band Data instrument (WBD), the Plasma Electron And Current Experiment (PEACE), the

Diameter	2.9 m
Height	1.3 m
Mass	1200 kg
Scientific payload	71 kg
Solar array power	224 W
Spin rate	15 rpm
Lifetime	5 years

Table 6.1: *Spacecraft statistics of the Cluster satellites (from ESA Cluster homepage).*

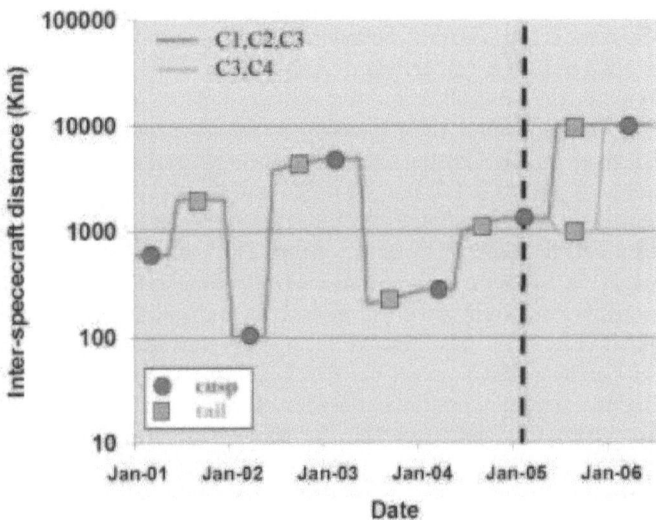

Figure 6.1: *Distance between the spacecraft during the Cluster mission (from ESA Cluster homepage).*

Cluster Ion Spectrometry experiment (CIS), and the Research with Adaptive Particle Imaging Detectors (RAPID). For the purpose of this work, the magnetic field measurements of the FGM are essential, but also CIS data are used in order to determine the plasma velocity associated with NFTEs.
An advantageous situation to analyze NFTEs occurs when the satellites have a large separation, since we can use the measurements of each satellite to verify our method. Fig. 6.1 shows the different distances between the Cluster satellites during the last years. One can see that the largest separation was achieved from autumn 2002 until autumn 2003, and from summer 2005. Therefore, the events studied in the following will be from these two periods.

6.2 Substorm and a series of NFTEs on September 8th 2002

On September 8th, 2002, an isolated substorm with a peak AE of about 400 nT occurred between 20 and 23 h UT. A favorable constellation of multiple spacecraft and ground observations allowed to reconstruct in details the time sequence of this substorm and to model the near-Earth magnetic configuration, see Sergeev et al. (2005). Following this description, a clear growth phase was observed after the arrival of a southward IMF after 20 UT. The auroral breakup, the intensification of a westward electrojet, and Pi2 pulsations consistently indicated the expansion phase onset at 21:18 UT in the 22–24 MLT sector. The Cluster tetrahedron was centered in the middle of magnetotail at $[-16.7; 0.2; 4.5]$ R_e GSM. The satellites exited from the thinning plasma sheet shortly after 21:00 UT, they were located outside of the plasma sheet at the time of interest. After 21:17 UT, a series of Earthward propagating 1 minute scale variations of the magnetic field and plasma flow components consistent with the picture of multiple NFTEs/flux ropes were observed (Fig. 6.2). The first NFTE appeared at about 21:17 UT and propagated Earthward at a speed V_x=625 km/s; V_y=-72 km/s (determined from timing of magnetic variations, Sergeev et al.,2005). After this NFTE, the plasma sheet continued to be thin for some 20 min until the transient plasma sheet expansions start to be observed. This is a favorable situation, because if the plasma sheet is thin, the approximation of a tangential discontinuity as an initial state is better justified. We applied our model to the NFTEs starting at 21:21 UT, 21:22:30 UT, and 21:24 UT, ignoring the first NFTE at 21:17 UT (in which the interaction of reconnected flux tube with previously closed plasma sheet flux tubes should be more pronounced than in the following development).

A comparison with the results of the theoretical model (Fig. 4.3) shows that the expected features for the perturbations are indeed found in the observations : the asymmetric bipolar variation of B_z, a compression of B_x, as well as a plasma flow v_z of cold O^+ ions directed to the plasma sheet are clearly visible. Also the change of the sign of B_z and v_z corresponds to the maximum in B_x. Therefore, we suppose that the observed perturbations can be treated in the frame of our theoretical model.

The GSM magnetic field data are obtained from the fluxgate magnetometer (FGM) experiment (Balogh et al., 2001) with 1 s time resolution. The O^+ moments with 4 s time resolution were measured by the Composition and Distribution Function Analyser (CODIF) of the Cluster Ion Spectrometry (CIS) experiment (Rème et al., 2001) observed at the Cluster spacecrafts. The O^+ data was only used if the O^+ density exceeded 0.005 cm^{-3}. To evaluate the integration kernel $K(p)$ we also need to know the spacecraft location with respect to the reconnection site (Figure 5.8). Fortunately, the actual z–position of the neutral sheet (about +1 R_e in z–direction) is known from the modelling made in Sergeev et al. (2005). Therefore, the z–distance between the satellite and the reconnection site is approximately 3.5 R_e. The local velocity of the disturbances is determined by using multipoint timing analysis

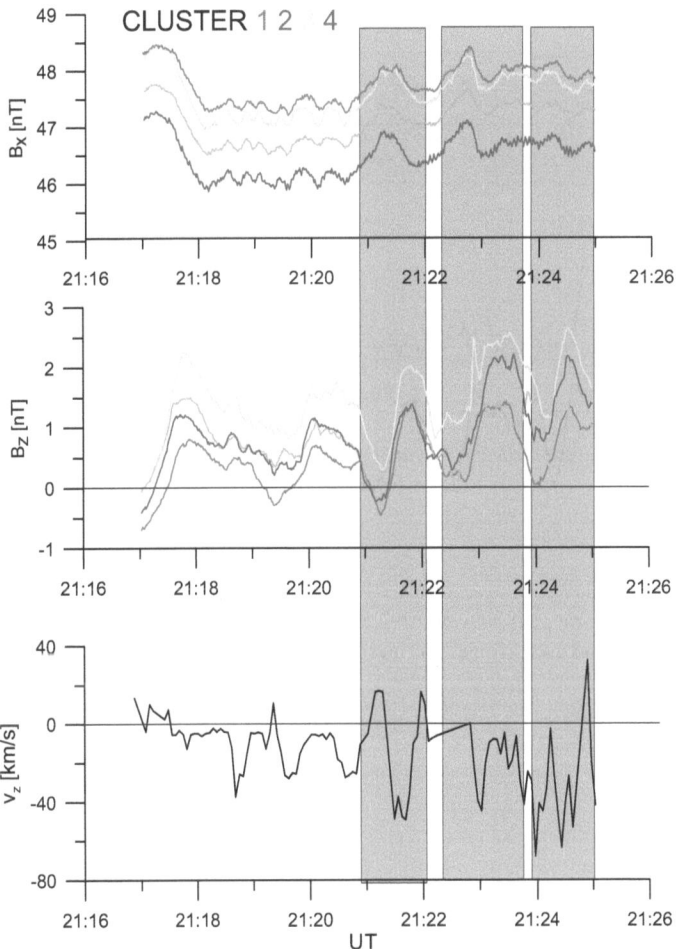

Figure 6.2: *Event on September 8th, 2002, observed by the four Cluster satellites. We analyze the NFTEs starting at 21:21 UT, at 21:22:30 UT and 21:24 UT (shaded areas).*

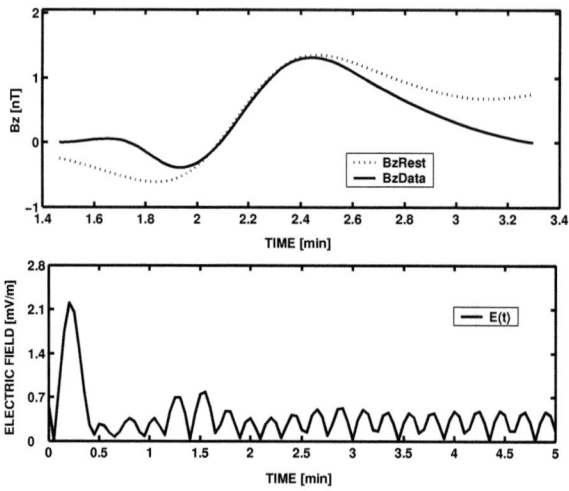

Figure 6.3: *Behavior of the reconnection electric field for NFTE starting at 21:21 UT (lower panel), and initial (solid line) and reconstructed magnetic field perturbations (dashed line) using C1.*

(e.g., Harvey, 1998), giving about 700 km/s. We assume that this velocity is approximately the Alfvén velocity. Since we know the Alfvén velocity, it is possible to introduce dimensional quantities. The background magnetic field is $B^{(0)} = 50$ nT, the characteristic time scale is $T^{(0)} = 60$ s, and the characteristic velocity is $v_A = 700$ km/s. This gives a characteristic length scale of $L = v_a T^{(0)} = 42000$ km ≈ 6.6 R_e. This means that one normalized length unit used in the theoretical description in the previous sections corresponds to 6.6 R_e for this special case.

Application of our model to the NFTE starting at 21:21 UT in Fig. 6.3 leads to a reconnection electric field of 2.1 mV/m over a time period of about 30 s (Fig. 6.3) from C1. The location in x–direction was found to be 12.5 R_e tailwards of the satellite, corresponding to a location of the reconnection site at 29.2 R_e in the magnetotail. From C2 a reconnection rate of 1.5 mV/m and a distance to the reconnection site is 13.2 R_e, giving a distance to Earth of 29.9 R_e. C3 gives a reconnection rate of 1.6 mV/m at a distance of 14 R_e. This implies a reconnection site at 30.7 R_e in the magnetotail. Finally, from C4 a reconnection rate of 1.6 mV/m at a x–distance of 14.2 R_e from the satellite, corresponding to a reconnection site 30.9 R_e tailwards of the Earth. Recapitulating, the reconnection rate is found to be between 1.5 and

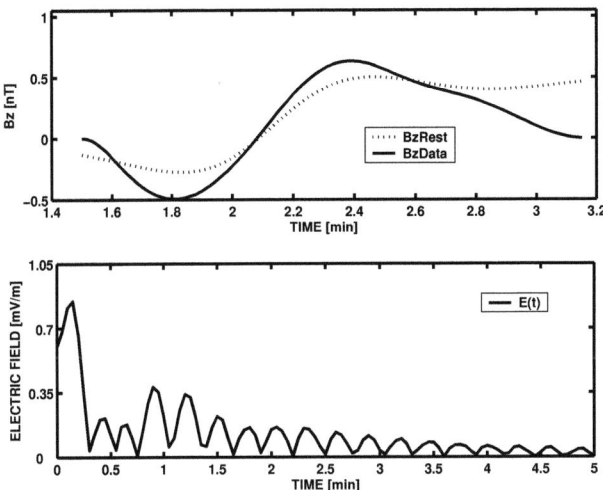

Figure 6.4: *Behavior of the reconnection electric field for NFTE starting at 21:22:30 UT (lower panel), and initial (solid line) and reconstructed magnetic field perturbations (dashed line) using C1.*

2.1 mV/m, while the location of the reconnection site is between 29.2 and 30.9 R_e in the magnetotail.
One can see that the reconstructed amplitude of the reconnection electric field and the reconnection site are consistent among the four satellites. For example, C3 is located about 0.5 Re lower than C2, therefore the observed amplitude in the magnetic field perturbation at C3 is 1.7 nT compared with 1.3 nT at C2. Yet, the reconstructed amplitude of the electric field is quite similar, i.e., 1.6 mV/m from C3 and 1.5 mV/m from C2. Therefore our model shows that the observed features at all four satellites are associated with the same reconnection event.

The reconstructed electric field for the NFTE at 21:22:30 UT from C1 is 0.9 mV/m with a time duration of about 20 s (Fig. 6.4). Again the location of the reconnection site is at 29.2 R_e in the magnetotail. C2 gives a reconnection rate of 1.4 mV/m and at a distance of 13.3 R_e, giving the location of the reconnection site at 30 R_e. We do not use C3 for the reconstruction, because the NFTE signal is not well pronounced. From C4, a reconnection rate of 1 mV/m at 12.8 R_e is found. The reconnection site is therefore located at 29.5 R_e in the magnetotail. Summarizing these results, the reconnection rate is between 0.9 and 1.4 mV/m and located between 29.2 and 30 R_e.

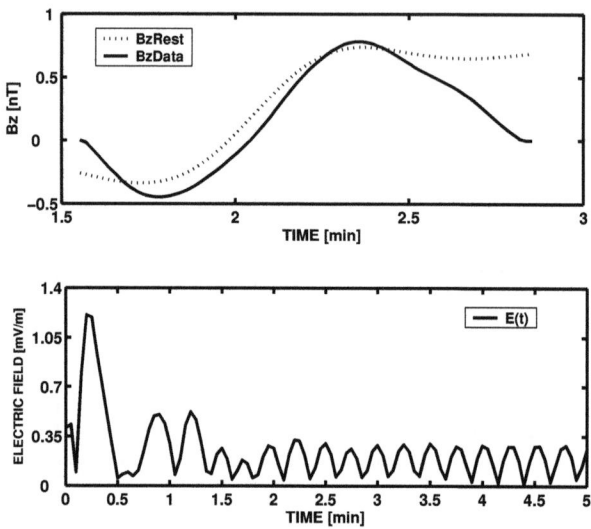

Figure 6.5: *Behavior of the reconnection electric field for NFTE starting at 21.24 UT (lower panel), and initial (solid line) and reconstructed magnetic field perturbations(dashed line) using C1.*

The reconstructed electric field for the NFTE starting at 21:24 UT from C1 is 1.2 mV/m with a time duration of about 30 s (Fig. 6.5). The reconnection process took place about 11.9 R_e tailwards from the satellite, giving the reconnection site at 28.6 R_e. From C2, the reconnection rate is found to be 1.4 mV/m, starting 12.6 R_e tailwards. Therefore, the reconnection site is located at 29.3 R_e. The reconnection rate obtained from C3 is 1.2 mV/m at the same location as found from C2. Finally, C4 gives the reconnection rate as 1.6 mV/m and a distance between the reconnection site and the satellite of 13 R_e. Thus, the reconnection site is at 29.7 R_e. From this event, the reconstructed reconnection rate is between 1.2 and 1.6 mV/m, and the reconnection site is at 28.6 to 29.7 R_e in the magnetotail.

The amplitude of the reconnection electric field is consistent with estimations of the magnetotail reconnection rate using ground–based measurements (Blanchard et al., 1996, 1997; Østgaard et al., 2005). The smaller pulses which can be seen in Figures 6.3 and 6.4 for times larger than 30 s are noise, resulting from the solution of the inverse problem. The amplitude of the noise mainly depends on the z–distance between the satellite and the reconnection structure (Semenov et al., 2005b). If the z–distance decreases, also

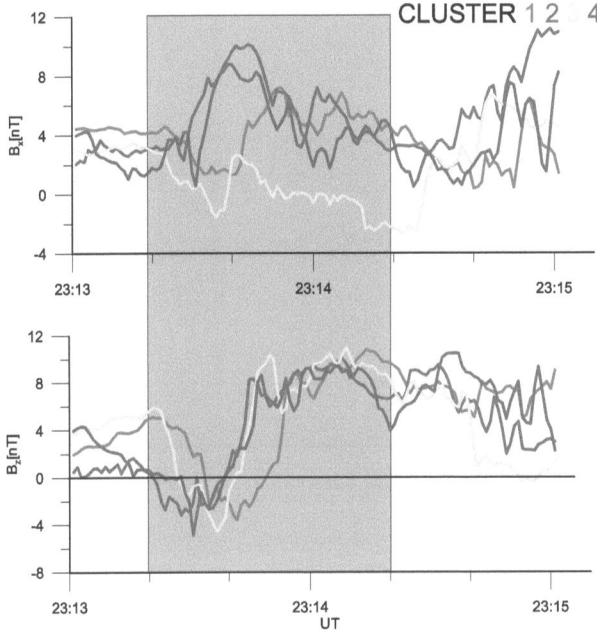

Figure 6.6: *Event on September 8th, 2002, observed by the four Cluster satellites. The shaded area indicates the analyzed NFTE.*

the amplitude of the noise is decreasing.

6.3 A NFTE on August 13th 2002 inside the plasma sheet

A substorm with a peak AE of about 400 nT occurred on August 13th, 2002, with electrojet intensification starting from about 22:52 UT (Penz et al., 2005). Cluster was located at the post–midnight sector at [-17.2; -6.9; 2-4] Re GSM and experienced thinning of the plasma sheet after 22:30 UT and exited into the lobe around 22:55 UT. Cluster then started to observe bursty flow enhancements with B$_z$ fluctuations as the spacecraft reencountered the plasma sheet at 23:06 UT. A series of Earth-ward flows and variations of the B$_z$ magnetic field similar to the September 8th events were detected except for the fact that Cluster was near the center of the plasma sheet within the current sheet as shown in the small B$_x$ values in Figure 6.6. In fact for this

73

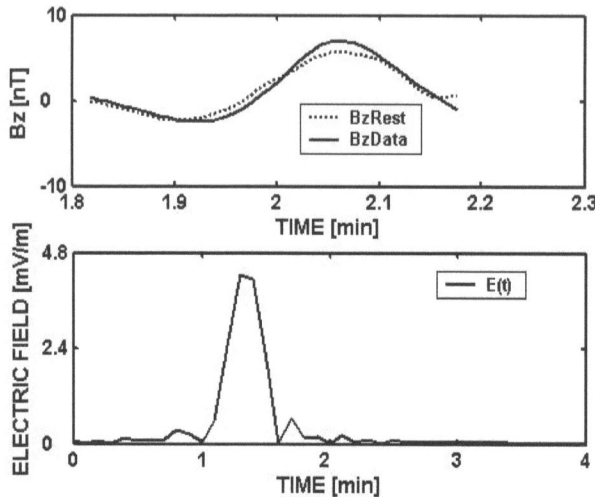

Figure 6.7: *Measured (solid line) and reconstructed magnetic field (dotted line) and the reconstructed reconnection electric field (lower panel) for the event on August 13th, 2002 from C1.*

particular event, B_x of C3 was almost zero, indicating that C3 was located at the center of the current sheet. For applying the analytical model we therefore use the z–distance from C3 as the location in z for each spacecraft, i.e., $(z_{C1}, z_{c2}, z_{c3}, z_{c4}) = (0.4, 0.5, 0.0, 0.4)$ Re. Note that if we simply fit the data into Harris current sheet assuming

- the lobe field value estimated from the total pressure at C4 between 23:13:20 and 23:14:20 UT, and
- C3 being at the center of the current sheet,

the thickness of the current sheet will be 1.3 Re. Hence, all the four spacecraft are located quite inside the plasma sheet. The reconstructed electric field for the NFTE at 21:13:20 UT from C1 is 4.2 mV/m with a time duration of about 40 s (Fig. 6.7). The location of the reconnection site is at 24.2 R_e in the magnetotail. C2 gives a reconnection rate of 4.6 mV/m and at a distance of 23.2 R_e. The reconnection rate from C3 is 4 mV/m at a location of 24.2 Re tailwards. From C4, a reconnection rate of 5 mV/m located at 24.7 R_e in the magnetotail. Summarizing these results, the reconnection rate is between 0.9 and 1.4 mV/m and located between 29.2 and 30 R_e. Summarizing, an reconnection electric field between 4 and 5 mV/m was found. The time duration of the pulse is in the range of 30 to 40 s, and the reconnection site is located 23.2 to 24.7 Re in the Earth magnetotail.

7 Extension to a compressible model

Considering Figs. 6.3 – 6.5, it is obvious that not all features of the magnetic field can be reconstructed quantitatively good, indicating that the plasma compressibility, which is neglected until now, should be implemented. One can see that the magnitude of the B_z component can be reconstructed well. This is because of the fact that the variations of the B_z component mainly depend on the inclination of the shock, which bounds the outflow region. Since the inclination is very similar for both the compressible and incompressible case, the incompressible model reproduces the B_z component satisfactory. Additionally, the incompressible model gives broader curves compared to the measurements. It will be shown in the following that a compressible model gives curves with a shorter time duration.

For the B_x component, the situation is somewhat different. Comparing the model results with the measured data, the amplitude of the variations of the B_x component are overestimated in the incompressible theory by a factor of at least 2. This is because the variations of B_x depend mainly on the size of the shocks bounding the outflow region, which will be smaller if the plasma is assumed to be compressible. Therefore, the incompressible model overestimates the variations in the B_x component. Therefore, a compressible theory for Petschek–type magnetic reconnection will be developed in this chapter.

Using the compressible method gives better results in several ways. The oscillations occurring in the incompressible model (e.g., Fig. 6.3) are suppressed. A much better correlation between the initial and reconstructed magnetic field perturbations is found for both satellite data and data generated by a direct model. And for multi–spacecraft observations, the reconstructed values are more consistent compared with the incompressible case.

7.1 MHD theory for a compressible plasma

After the local reconnection of magnetic flux according to the Petschek-type model of magnetic reconnection, the plasma and magnetic field behavior in the outer regions can be described within the frame of ideal compressible magnetohydrodynamics (Akhiezer et al., 1975; Semenov et al., 2004a),

$$\frac{\partial \rho}{\partial t} + \nabla \cdot (\rho \mathbf{v}) = 0, \qquad (7.1)$$

$$\frac{\partial (\rho \mathbf{v})}{\partial t} + \nabla \cdot \left[\rho \mathbf{v} \mathbf{v} + p \mathbf{I} - \frac{1}{4\pi} \left(\mathbf{B} \mathbf{B} - \frac{B^2}{2} \mathbf{I} \right) \right] = 0, \qquad (7.2)$$

$$\frac{\partial}{\partial t}\left(\frac{\rho v^2}{2}+\rho e+\frac{B^2}{8\pi}\right)+\nabla\cdot\left[\rho\mathbf{v}\left(\frac{v^2}{2}+e+\frac{p}{\rho}\right)+\frac{1}{4\pi}\mathbf{B}\times(\mathbf{v}\times\mathbf{B})\right]=0,$$
(7.3)

$$\frac{\partial\mathbf{B}}{\partial t}+\nabla\cdot(\mathbf{B}\mathbf{v}-\mathbf{v}\mathbf{B})=0.$$
(7.4)

These equations describe the conservation of mass (Eq. 7.1), momentum (Eq. 7.2), energy (Eq. 7.3), and magnetic flux (Eq. 7.4), respectively. ρ, $\mathbf{v}$, $\mathbf{B}$, p, and e are the mass density, the plasma velocity, the magnetic field, the isotropic pressure, and the internal energy, respectively, and $\mathbf{I}$ is the unit dyadic.

These MHD equations can be linearized with respect to constant background quantities, in the following denoted by superscript (0). The linearization allows for the description of all first order quantities (indicates by superscript (1)) in terms of a displacement vector $\boldsymbol{\xi}$,

$$\mathbf{v}^{(1)}=\left(\frac{\partial}{\partial t}+\mathbf{v}^{(0)}\cdot\nabla\right)\boldsymbol{\xi},$$
(7.5)

$$\mathbf{B}^{(1)}=\mathbf{B}^{(0)}\cdot\nabla\boldsymbol{\xi}-\mathbf{B}^{(0)}\nabla\cdot\boldsymbol{\xi},$$
(7.6)

$$\rho^{(1)}=-\rho^{(0)}\nabla\cdot\boldsymbol{\xi},$$
(7.7)

$$p^{(1)}=c_S^2\rho^{(1)},$$
(7.8)

$$P^{(1)}=-\rho^{(0)}\left[u^2\nabla\cdot\boldsymbol{\xi}+(\mathbf{v}_A\cdot\nabla)\mathbf{v}_A\cdot\boldsymbol{\xi}\right].$$
(7.9)

Here, $P^{(1)}$ is the first order total pressure, while $\mathbf{v}_A$ and c_S are the Alfvén velocity and the sound speed, respectively, and $u^2=v_A^2+c_S^2$.

Substituting these expressions (Eq. 7.5–7.9) into the linearized MHD equations, applying a Laplace transformation with respect to time t, and a Fourier transformation with respect to the spatial coordinate x, the z-component of the displacement vector $\xi_z(p,k,z)$ fulfills the ordinary differential equation

$$\frac{\partial^2}{\partial z^2}\xi_z - q^2(p,k)\xi_z = 0,$$
(7.10)

with

$$q^2 = \frac{p^4+u^2k^2p^2+v_A^2c_S^2k^4}{p^2u^2+v_A^2c_S^2k^2}.$$
(7.11)

The solutions of Eq. 7.10 which satisfy first order total pressure balance at $z=0$ and vanish at infinity are

$$\xi_z(p,k,z)=\frac{\tilde{L}}{L+\tilde{L}}Q(p,k)e^{-qz} \qquad z>0,$$
(7.12)

$$\tilde{\xi}_z(p,k,z) = -\frac{L}{L+\tilde{L}}Q(p,k)e^{qz} \qquad z<0, \qquad (7.13)$$

where

$$L(p,k) = -\rho^{(0)}\frac{p^2 + v_A^2 k^2}{q(p,k)}. \qquad (7.14)$$

Tilde indicates the same function evaluated for the lower half space. Because of the finite boundary layer at $z=0$, the source function $Q(p,k) = \xi_z(p,k,0) - \tilde{\xi}_z(p,k,0)$ is non-zero and can be found from the solution of the Riemann problem (see Appendix). Summing up Eqs. A.32–A.36, one gets the source function as

$$\begin{aligned}Q(t,x,y) &= \Phi_1(\mathbf{w}_A^{(0)}) - \Phi_0(\mathbf{w}_A^{(0)}) + \Phi_2(\mathbf{w}_S^{(0)}) - \Phi_1(\mathbf{w}_S^{(0)}) \\ &+ \tilde{\Phi}_1(\tilde{\mathbf{w}}_S^{(0)}) - \tilde{\Phi}_2(\tilde{\mathbf{w}}_S^{(0)}) + \tilde{\Phi}_0(\tilde{\mathbf{w}}_A^{(0)}) - \tilde{\Phi}_1(\tilde{\mathbf{w}}_A^{(0)}).\end{aligned} \qquad (7.15)$$

The de Hoffmann-Teller velocities $\mathbf{w}^{(0)}$ are different for the $x>0$ and $x<0$ half spaces because the waves propagate in opposite direction away from the reconnection line. Hence, the Laplace-Fourier transform has to be done separately for $x>0$ and $x<0$ which doubles the number of terms in (7.15). Note, that a minus sign has to be taken in the Laplace–Fourier transform of the shifted Φ function (A.37) for $x<0$

$$\mathcal{L}_t\mathcal{F}_{xy}\left\{\Phi\left(t-\frac{x}{w_x}, y-\frac{w_y}{w_x}x\right)\right\} = -\frac{w_x \Phi(p, k_y)}{p + i\mathbf{w}\cdot\mathbf{k}}, \qquad x<0. \qquad (7.16)$$

Using this representation, after some algebra the expression for the source function is found as

$$\begin{aligned}Q(s) &= \left(\frac{1}{B_{1a}} - \frac{1}{B_a}\right)\frac{w_{Aa}^+}{1+isw_{Aa}^+} + \left(\frac{1}{B_{2a}} - \frac{1}{B_{1a}}\right)\frac{w_{Sa}^+}{1+isw_{Sa}^+} \\ &+ \left(\frac{1}{B_{\tilde{1}b}} - \frac{1}{B_{\tilde{2}b}}\right)\frac{w_{Sb}^+}{1+isw_{Sb}^+} + \left(\frac{1}{B_b} - \frac{1}{B_{\tilde{1}b}}\right)\frac{w_{Ab}^+}{1+isw_{Ab}^+} \\ &- \left(\frac{1}{B_{1a}} - \frac{1}{B_a}\right)\frac{w_{Aa}^-}{1+isw_{Aa}^-} - \left(\frac{1}{B_{2a}} - \frac{1}{B_{1a}}\right)\frac{w_{Sa}^-}{1+isw_{Sa}^-} \\ &- \left(\frac{1}{B_{1a}} - \frac{1}{B_{\tilde{2}b}}\right)\frac{w_{Sb}^-}{1+isw_{Sb}^-} - \left(\frac{1}{B_b} - \frac{1}{B_{\tilde{1}b}}\right)\frac{w_{Ab}^-}{1+isw_{Ab}^-}. \end{aligned} \qquad (7.17)$$

Here, w_A and w_S denote the velocity of the Alfvén discontinuity and the slow shock. The number indicate the different regions according to the Riemann problem as follows: 1–A_aS_a, 2–S_aC, $\tilde{2}$–CS_b, $\tilde{1}$–S_bA_b, while superscripts "+,−" indicate the regions $x>0$ and $x<0$, respectively. In the two–dimensional case, the source function $Q(s)$ consists only of six terms (Semenov et al.,

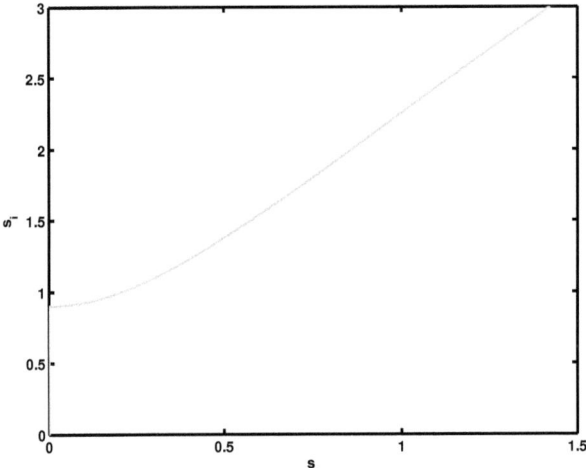

Figure 7.1: *Cagniard contour for the compressible case. Compared with the incompressible case, where the Cagniard contour was given explicitly, in the compressible case it has to be found from solving a fourth-order polynomial.*

1998), because the Alfvén discontinuity develops only in the region with higher Alfvén velocity.

To get the magnetic field and other physical quantities the inversion of the displacement vector to time–coordinate space is necessary. The inverse Fourier transformation gives

$$\xi_z(p,x,z) = \frac{1}{2\pi} \int_{-\infty}^{\infty} \zeta(p,k,z) e^{ikz} dk = \frac{1}{\pi} \int_0^{\infty} \Re \frac{\tilde{L}}{L+\tilde{L}} Q(s) F(p) e^{-p\tau(s)} ds, \qquad (7.18)$$

where the reality of the function $\xi_z(p,x,z)$ is taken into account. Compared with the incompressible case, the variable $\tau(s) = zq(s) - isx$ is now a fourth-order polynomial in s. A variable $s = k/p$ is introduced, and $F(p)$ is the reconnected flux. The advantage of the new variable is that it is possible to apply the shift theorem of the Laplace transformation $\mathcal{L}^{-1}\left(e^{-p\tau}F(p)\right) = F(t-\tau)$ if the contour can be analytically deformed in the complex s-plane in such a way that $\tau(s)$ becomes real along it. Thus, the inverse Laplace transformation is performed analytically. The function $s(\tau)$ can not be written explicitly; the analytical expression is available only for the inverse function $\tau(s)$. Therefore, actually the integration is carried out along the complex contour C such that $\tau(s)$ stays real (Fig. 7.1) for $0 \leq s \leq s_{max}$. The resulting time–coordinate representation of the displacement vector is

$$\xi_z(x,z,t) = \frac{1}{\pi}\int_C \Re \frac{\tilde{L}}{L+\tilde{L}} Q(s) F(t-\tau(s)) ds,, \qquad (7.19)$$

The source function $Q(s)$ includes all discontinuities and shocks which form the BL–structure and act as sources of perturbations.

Since the displacement vector is known, the derivation of the MHD quantities is straight forward and equal to the derivation done for the incompressible case. As an example, the variation of the B_z component is given as (Penz et al., 2006b)

$$B_z^{(1)}(x,z,t) = -B\frac{\partial}{\partial z}\xi_z = \frac{B}{\pi}\int_C \Re \frac{\tilde{L}}{L+\tilde{L}} i\, s\, Q(s)\, E(t-\tau(s))\, ds. \qquad (7.20)$$

A transformation to the variable τ is done, where

$$ds = \frac{1}{\tau_s} d\tau, \qquad (7.21)$$

and τ_s indicates a derivative of τ with respect to s. This gives the z–component of the magnetic field as the convolution integral

$$B_z^{(1)}(x,z,t) = \frac{B}{\pi}\int_0^t \Re \frac{\tilde{L}}{L+\tilde{L}} i\, s\, Q(s)\, E(t-\tau(s)) \frac{1}{\tau_s}\, d\tau. \qquad (7.22)$$

The end point of integration is defined from the condition of causality $\tau(s_{max}) = t$. The expressions for the other quantities have a similar mathematical structure.

We can now compare magnetic field trajectories calculated using the incompressible and the compressible model (Fig. 7.2). From this comparison it is possible to derive some tendencies. One can see that the magnitude of the B_z component remains approximately the same. This is because of the fact that the variations of the B_z component mainly depend on the inclination of the shock, which bounds the outflow region. Since the inclination is very similar for both the compressible and incompressible case, the incompressible model reproduces the B_z component satisfactory. Additionally, we see that the duration of the features in compressible plasma is getting shorter. This is favorable, because in Figures 6.3 and 6.4 it can be seen that the incompressible model gives broader curves compared to the measurements.

For the B_x component, the situation is somewhat different. From the middle panel of Figure 7.2 it is evident that the amplitude of the variations of the B_x

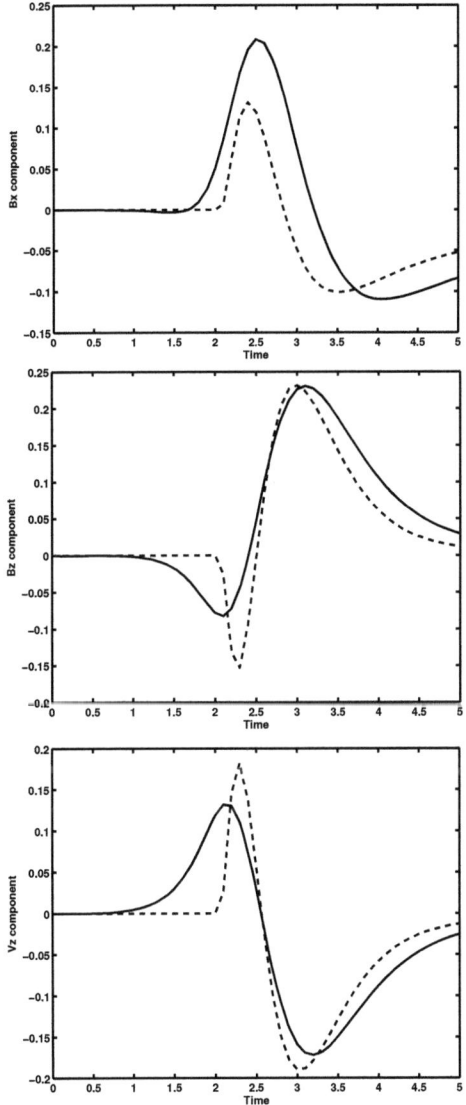

Figure 7.2: *Comparison of the x–component (upper panel) and the z–component (middle panel) of the magnetic field and the z–component of the plasma velocity (lower panel) for $z = 0.5$ for incompressible plasma (solid line) and compressible plasma with $\beta = 0.1$ (dashed line).*

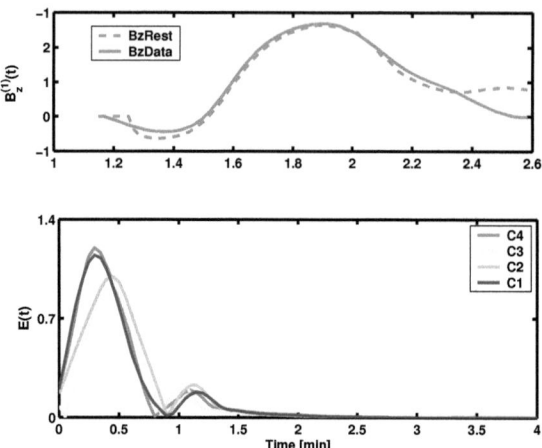

Figure 7.3: *Reconstructed reconnection rate from the four Cluster satellites, which has an amplitude of 1–1.1 mV/m and a duration of about 50 s (lower panel). The upper panel shows a comparison between the measured $B_z^{(1)}$ and the restored $B_z^{(1)}$ from the reconstructed reconnection rate for C4.*

component are overestimated in the incompressible theory by a factor of 2, which we also found by comparing the model results with the measured data. This is because the variations of B_x depend mainly on the size of the shocks bounding the outflow region, which will be smaller if the plasma is assumed to be compressible. Therefore, the incompressible model overestimates the variations in the B_x component. For the z–component of the plasma flow, the upper panel of Figure 7.2 shows that in compressible plasma, the amplitude is slightly increasing, while the duration is getting slightly shorter.

7.2 Application to the NFTEs on September 8th, 2002

As a first task, we apply the compressible model to the same event as the incompressible model in order to compare the results of both methods. The reconnection electric field reconstructed for the NFTE starting at 21:21 UT in Fig. 6.2 is 1.1 mV/m over a time period of about 50 s from C1. C2 and C3 give a reconnection rate of 1.0 mV/m, while C4 gives 1.1 mV/m (Fig. 7.3). The distance x between C1 and the reconnection site is 8.6 R_e, so that the reconnection site is located at 24.7 R_e in the magnetotail (Fig.

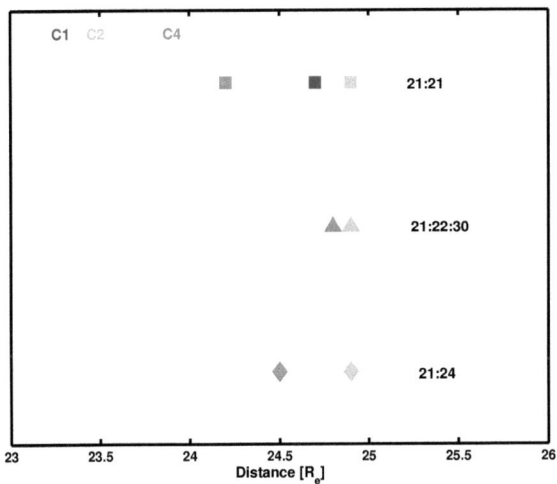

Figure 7.4: *Location of the reconnection site reconstructed for the NFTE starting at 21:21 UT (square), 21:22:30 UT (triangle), and 21:24 UT (diamond) from the Cluster satellites.*

7.4). For C2, $x = 8.6$, giving 24.9 R_e as the location of the reconnection site. From C3 x is found to be 8.4, giving the reconnection site at 24.6 R_e. C4 gives $x = 7.6$, corresponding to a reconnection site at 24.3 R_e. One can see that the reconstructed amplitude of the reconnection electric field and the reconnection site are consistent among the four satellites. For example, C3 is located about 0.5 Re lower than C2, therefore the observed amplitude in the magnetic field perturbation at C3 is 1.7 nT compared with 1.3 nT at C2. Yet, the reconstructed amplitude of the electric field is quite similar, i.e., 1.0 mV/m from C3 and from C2. The same can be found for the x–distance, where C4 is located closest to the reconnection site, and therefore the model gives the smallest distance between the satellite and the reconnection site. Therefore our model shows that the observed features at all four satellites are associated with the same reconnection event.

Analysis of the event starting at 21:22:30 UT was done only with C2 and C4, since the other two satellites did not observe pronounced NFTE–like signatures. The reconnection rates are found to be 0.95 and 1.0 mV/m for C2 and C4, respectively. C2 gives $x = 8.6$, leading to a distance of 24.9 R_e for the reconnection site. From C4, x is found to be 8.25 R_e, giving a distance of 24.8 R_e in the magnetotail. For the event starting at 21:24 UT, data from

C2, C3, and C4 are used. The reconnection rate is found to be 0.95 mV/m (C2), 0.91 mV/m (C3), and 0.9 mV/m (C4). According to these data, the reconnection site is located at 24.9 R_e (C2), 23.9 R_e (C3), and 24.5 R_e (C4) in the magnetotail.

As already predicted (Semenov et al., 2005a), the reconnection rate is smaller than inferred from the incompressible model. Also the distance between the reconnection site and the satellite decreased, while the duration of the pulses increase. All these features are consistent with the qualitative estimates done by Semenov et al. (2005a). However, a significant reduction of the noise resulting from the solution of the inverse problem is achieved compared with the incompressible model. Also the variations of the reconstructed reconnection rate and site between the satellites decreased.

7.3 NFTE on September, 26th, 2005

Another event showing pronounced FTEs was observed by Cluster on September, 26th, 2005 (Fig. 7.5). The advantage of this events is the fact that tailward propagating NFTEs are observed at 8:43 and at 9:00 UT. In contrast to the NFTEs on September 8th, 2002, where the FTEs are moving Earthward, in the case of tailward travelling NFTEs, we can restricted the possible location of the reconnection site. It should be located between the region where the dipole structure of the Earth's magnetosphere plays a dominant role over the stretched field line structure of the magnetotail. This transition region is approximately located at 10 R_e. Since the Cluster satellites are located at about 15 R_e, the reconnection site must be located somewhere in the region between 10 and 15 R_e. Another indication that the satellites are located close to the reconnection site is that the bipolar variation of the B_z component is not well established, meaning that the distance between the reconnection site and the satellites is small.

In the following, the NFTE starting at about 8:43 UT is analyzed. From timing analysis, an Alfvén velocity of 900 km/s is found. The z-distance of C2, which is the closest satellite to the current sheet, is assumed to be 2000 km. C1 was located at -16.0 R_e and had a z-distance of 0.078 normalized units (Normalization parameters: $B_0 = 40$ nT, $v_A = 900$ km/s, $T_0 = 60$ s), corresponding to 4200 km above the current sheet. At C2, the NFTE started at 8:42:43 UT. The reconstructed reconnection electric field is 6.8 mV/m, and the distance between the reconnection site and the satellite is found to be 2 R_e. C2 was located nearly at the same x-distance as C1, namely at -16.1 R_e. Thus, the NFTE starts at 8:42:41 UT. The normalized z-distance is 0.04, corresponding to 2000 km. The reconstruction gives a reconnection electric

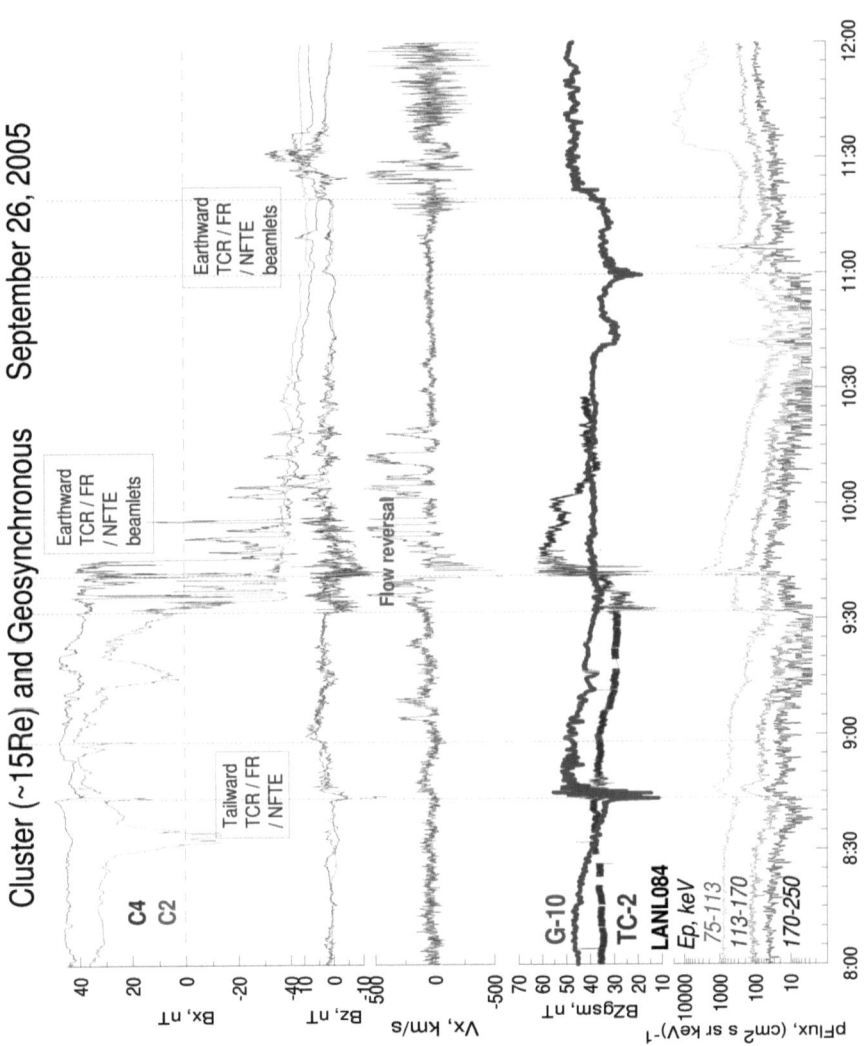

Figure 7.5: *Event on September, 26th, 2005, seen from different spacecrafts. Earth- and tailward moving NFTEs are observed. We analyze the event starting at 8:43 UT (Courtesy of V. A. Sergeev).*

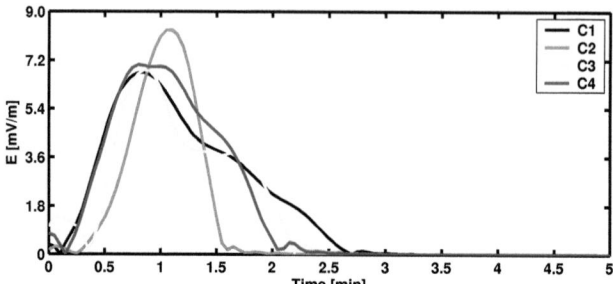

Figure 7.6: *Reconstructed electric field for the event on September, 26th, 2005. The time duration is about 2 min and the amplitude of the reconnection electric field is between 6.8 and 8.3 mV/m.*

field of 8.3 mV/m and a distance of 2 R_e. C3 was located close to Earth than C1 and C2 with a distance of -14.6 R_e. Therefore, C3 observed the NFTE much earlier at 8:42:32 UT. The z–distance between C3 and the current sheet was 0.76 normalized units (4850 km). The reconstructed electric field was found as 6.8 mV/m, at a distance of 0.8 R_e away from the satellite. C4 was located at a similar x–distance as C3 (-14.5 R_e), but slightly further away in z–distance (0.94 or 6000 km). The signature of the NFTE arrived at C4 at 8:42:37 UT. Application of the reconstruction method gives a reconnection rate of 7 mV/m at a distance of 0.8 R_e.

A summary of the results is shown in Figs. 7.6 and 7.7. Fig 7.6 shows the reconstructed reconnection electric field from all four Cluster satellites. A good agreement in the amplitude as well as in the time duration can be seen. The time duration is about 2 min, however C2 gives a slightly smaller time. This is because for C2 the end of the NFTE signature is not clearly expressed, therefore it is difficult to determine the time duration of the reconnection process from C2. Additionally, C2 is closest to the current sheet, which may also influence the result of the reconnection method. In Fig. 7.7 the location of the four satellites (circles) and of the reconstructed reconnection site (diamonds) is shown. Despite of the different locations of the satellites, the position of the reconstructed reconnection site agrees for all four satellites. This indicates that the reconnection process was initiated slightly before 8:42:30 UT at a location somewhere around 14 R_e.

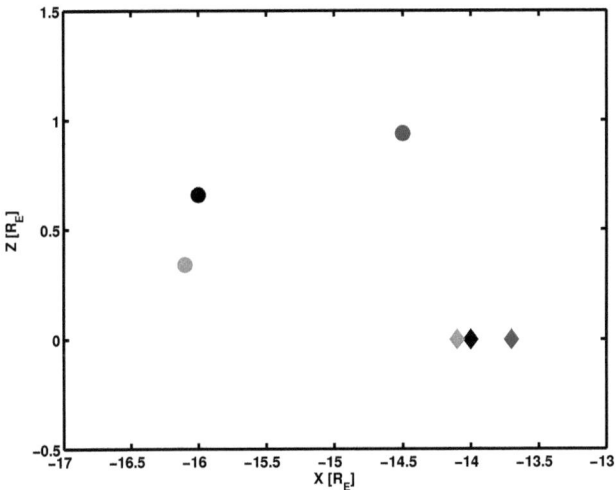

Figure 7.7: *Location of the satellites (circles) and the reconnection site (diamonds). The reconnection site is located between -13.7 and -14.1 R_e in the tail.*

7.4 Analysis of high latitude FTEs

Two periods of sustained FTE activity observed by the Cluster satellites are investigated in the following. On January 21st, 2001, a series of FTEs are observed at the poleward edge of the cusp around 15 MLT (Fig. 7.8). Similar features are observed on February 12th, 2001, but located at 13:30 - 14:00 MLT at the high-latitude magnetopause at the equatorward edge of the cusp (Fig. 7.8). In both cases radar signatures of transient reconnection were studied (Wild et al., 2001; Farrugia et al., 2004). Thus, the presented method gives an independent estimate of the reconnection rate compared to the radar signatures. Additionally, Pinnock et al. (2003), studying the reconnection rate from the ionospheric viewpoint for an interval of steady reconnection during southward IMF, obtained its variation as a function of magnetic local time. Significantly, they obtained a local minimum around 14 MLT and a sharp increase at later times. Using the reconstruction method it is possible to confirm these observations.

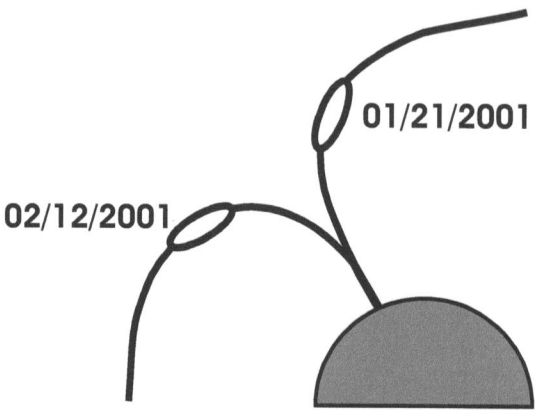

Figure 7.8: *Two periods of FTEs at the high–latitude magnetopause studied. On January 21st, 2001, the FTEs are observed poleward of the cusp, while on February 12th, 2001, they are located equatorward.*

7.4.1 FTEs on 14th February 2001

We analyze a series of FTEs at the high–latitude magnetopause on 14 February 2001 observed by the Cluster satellites, which were already discussed by Wild et al. (2001). The transformation from GSM coordinates to boundary normal coordinates was done by using a unit normal vector $bfn = (0.669, -0.262, 0.696)$ (Wild et al., 2001). At least 8 FTEs can be observed in the time interval lasting from 9:30 to 11:15 UT (Fig. 7.9). At the beginning of the interval shown in Fig. 7.9, the spacecraft was located within the magnetosphere where the field strength was about 20 nT and directed principally southward and westward. At the end of the interval, the spacecraft was located in the magnetosheath where the field was still about 20 nT, but much more rapidly varying, and directed southward and eastward. Three principal magnetopause boundary crossings have been identified, occurring at 10:17, 10:22, and 10:33 UT. During the initial magnetospheric interval, prior to the first magnetopause encounter, the field became increasingly variable in both direction and magnitude as the magnetopause was approached. These variations appear to have begun near 9:00 UT. Prior to the first magnetopause encounter, four clear magnetospheric FTEs are observed, centered near 9:45, 9:54, 9:59, and 10:04 UT. After the observation of the four magnetospheric FTEs, the spacecraft entered a magnetopause boundary layer at 10:09 UT, which endured until the first magnetopause crossing at 10:17 UT. Following the transition into the magnetosheath at 10:17 UT, the spacecraft then briefly re-entered the magnetosphere between 10:22 and 10:33 UT. In the following traversal of the magnetosheath, FTEs were again observed. Two prominent examples are centered near 10:46 and 11:01 UT, with events of smaller amplitude occurring at 10:36 and 10:43 UT.

First, we analyze the four FTEs, where the satellites were still located inside the magnetosphere. We applied the inverse model to all four FTEs indicated by the dashed lines in Fig. 7.11. The Alfvén velocity is about 500 km/s, while the normal distance to the magnetopause is found from a model developed by Shue et al. (1997) (Fig. 7.10). The distances used in the following are 1 R_E for FTE1, 0.7 R_E for FTE2 and FTE3, and 0.5 R_E for FTE4. In Fig. 7.11 the results of the inverse model are shown. FTE1 is caused by an reconnection electric field of about 4 mV/m at a site 3.5 R_E away from the satellite with a time duration of about 90 s. FTE2 corresponds to a reconnection electric field of 2.4 mV/m with a duration of nearly 2 min at a distance of about 3.5 R_E. A similar electric field of 2.3 mV/m at a distance of 3.5 R_E is found for FTE3, but the time duration of 60 s is smaller compared with the previous FTEs. The last magnetosphere FTE, FTE4, gives the smallest electric field of 1.8 mV/m with a duration of 2 min, originated 4 R_E away from the satellite. It is possible to derive some tendencies from Fig. 7.11. FTE1 is furthest away from the magnetopause, but has an amplitude of about 20 nT comparable to the other three FTEs, therefore it is obvious that FTE1 gives the largest electric field amplitude. FTE3 has a shorter duration than the other three, implying that also the reconnection pulse is shorter compared with the others.

After several crossings of the magnetopause, Cluster 4 observed four additional FTEs, located in the magnetosheath. Again the normal distance is inferred from Fig. 7.10. We achieve a distance of 1 R_E for FTE5 and FTE6, 0.6 R_E for FTE7, and 0.8 R_E for FTE8. In Figure 7.12, the reconstructed electric field for these FTEs is shown. FTE5 is not clearly expressed, therefore the restored electric field is only 0.9 mV/m. The time duration is slightly below 1 min, while the reconnection site is located 3.5 R_E away from the reconnection site. FTE6 gives a reconnection electric field of 1.3 mV/m over 1 min with a reconnection site again 3.5 R_E away. FTE7 is better expressed than the previous magnetosheath FTEs, giving a electric field 3.4 mV/m over 80 s. The reconnection site is located about 2.8 R_E. A similar result is found for FTE8, but the time duration is more than 2 min for this event. By comparing B_N and the reconnection electric field, one can again see a good correlation in the time duration, the amplitude of the FTEs and the reconstructed electric field.

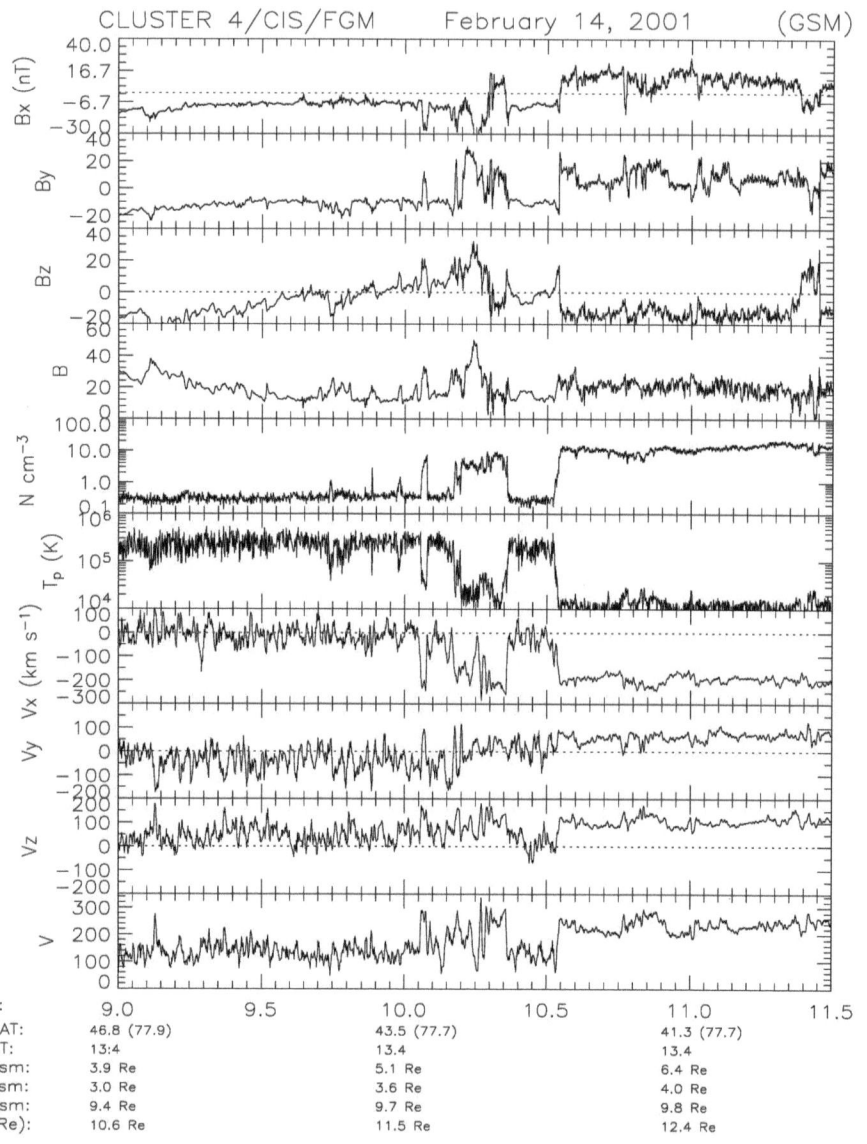

Figure 7.9: *Event on February 14th, 2001, seen from C4. Magnetospheric FTES are identified at 9:45, 9:54, 9:59, and 10:04 UT, while magnetosheath FTEs can be seen at 10:36, 10:43, 10:46, and 11:01 UT (after Wild et al., 2001).*

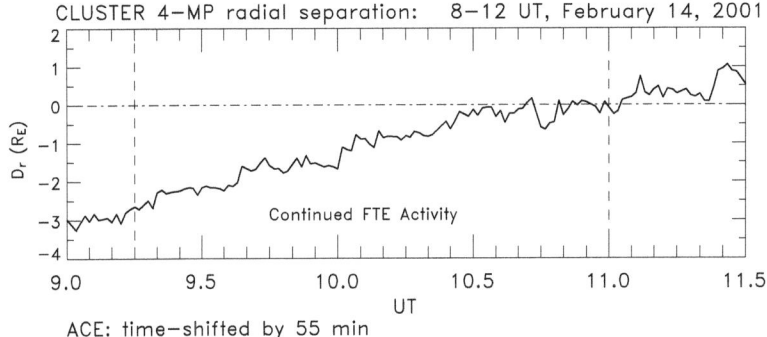

Figure 7.10: *Normal distance between the magnetopause and C4 for the considered interval.*

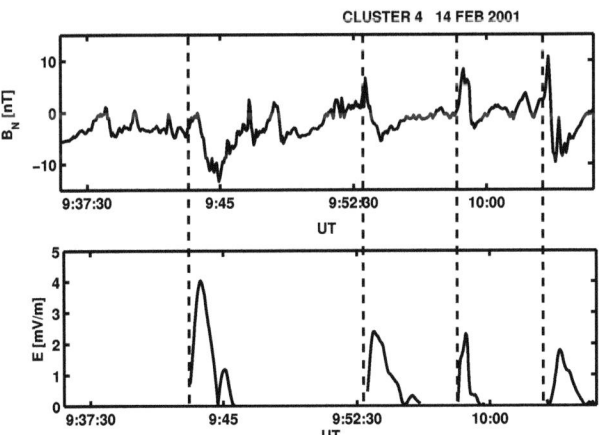

Figure 7.11: *Four magnetosphere FTEs observed by C4 and the reconstructed reconnection electric field.*

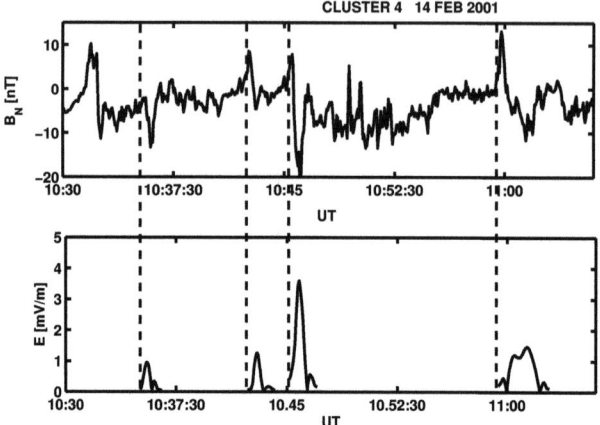

Figure 7.12: *Four magnetosheath FTEs observed by C4 and the reconstructed reconnection electric field.*

7.4.2 FTEs on 21th January 2001

On 21th January 2001, Cluster observed several FTE–like signatures during a time interval lasting from 15:30 to 16:00 UT (Farrugia et al., 2004). Very low values of the density and a fluctuation–free magnetic field indicate that the spacecraft is inside the magnetosphere until 14:10 UT. From 14:10 to 15:30 UT, the high density and sporadically depressed field indicate a traversal through the exterior cusp. The positive B_x indicates further that the spacecraft is tailward of the bifurcation line in the Earth's magnetic field. A reorientation of the field during the passage of the high pressure solar wind takes place during 15:20-15:30 UT, where $B_z \sim 0$ and the field is mainly in B_x (> 0) and B_y. This is possibly an encounter with the northern tail lobe. At 15:35 UT when the solar wind dynamic pressure decrease occurs, and for the subsequent 25 min, Cluster 1 encounters a region of plasma characterized by bursts of high speed flows. Simultaneous with these sporadically enhanced flows, the magnetic field executes large-amplitude fluctuations. The positive GSM B_x and $B_z \sim 0$ strongly suggest that Cluster 1 is now situated in a boundary layer downsteam of the cusp. After 16:16 UT, the spacecraft enters the magnetosheath.

We choose four FTEs from this interval, occurring at 15:41, 15:43, 15:51, and 15:57 UT. We calculate the normal distant between the satellite and the magnetopause similar as in the previous case. The distance for C1 is shown

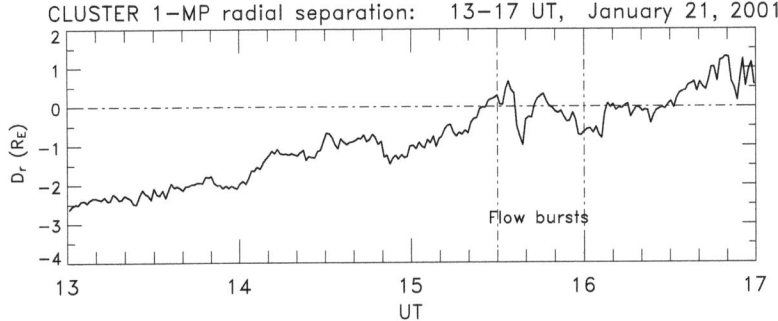

Figure 7.13: *Normal distance between the magnetopause and C1.*

in Fig. 7.13. The local Alfvén velocity is found to be 250 km/s. We determined the normal vector from a cross product of the magnetospheric and the magnetosheath magnetic field, giving $\mathbf{n} = (0.7946, 0.6041, 0.0639)$. Figure 7.14 shows the normal component of the magnetic field, the plasma flow velocity, and the reconstructed electric field. FTE1 gives a reconnection electric field of 4 mV/m over a time period of about 50 s. The distance between the satellite is estimated to be about 2.5 R_E for all four FTEs considered. The reconstructed electric field for FTE2 is 8 mV/m over about 30 s. FTE3 is the most expressed, giving a reconnection electric field of 11 mV/m over a time period of about 50 s. Finally, FTE4 corresponds to an electric field of 7.5 mV/m lasting about 50 s. As in the previous examples, also here we find a good correlation between the reconnection signatures and the electric field.

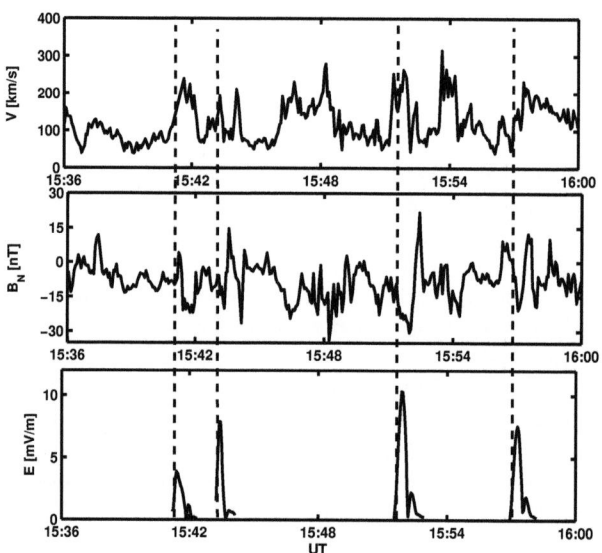

Figure 7.14: *Plasma flow velocity (upper panel), the B_N component (middle panel), and the reconstructed magnetic field (lower panel). The dashed lines indicate the four selected FTEs, showing a good correlation with flow bursts of the plasma velocity.*

8 Comparison between the analytical model and a numerical magnetotail simulation

Until now, we did not present a verification of the reconnection model. In order to do this, this section is devoted to a comparison between results of a numerical magnetotail simulation and the analytical method to reconstruct the reconnection rate by solving the inverse problem of Petschek-type magnetic reconnection. A numerical self-consistent solution of the compressible MHD equations using a TVD Lax-Wendroff scheme used to study reconnection-associated disturbances in a magnetotail-like plasma environment. Based on this simulation, time series for the magnetic field at different positions are derived. These time series are used as an input for the inverse method and the according reconnection electric field is calculated. A comparison between the electric field used in the numerical simulation and the reconstructed electric field shows good agreement. This investigation confirms that despite several simplifications used in the analytical inverse model, the results are consistent with the numerical simulation.

8.1 Simulation of non–stationary reconnection

8.1.1 Simulator description

We proceed with a description of the simulation technique. The plasma evolution is described in terms of compressible MHD, implying that no kinetic effects are taken into account. In this approach, a finite localized resistivity breaks the frozen–in constraint and sets up reconnection. The used MHD equations, written in conservative form, are

$$\frac{\partial \rho}{\partial t} + \nabla \cdot (\rho \mathbf{v}) = 0, \tag{8.1}$$

$$\frac{\partial \rho \mathbf{v}}{\partial t} + \nabla \cdot (\mathbf{v}\rho\mathbf{v} - \mathbf{BB}) + \nabla p_t = 0, \tag{8.2}$$

$$\frac{\partial \mathbf{B}}{\partial t} + \nabla \cdot (\mathbf{vB} - \mathbf{Bv}) = \mathbf{S_B}, \tag{8.3}$$

$$\frac{\partial e}{\partial t} + \nabla \cdot (\mathbf{v}e + \mathbf{v}p_t - \mathbf{BB} \cdot \mathbf{v}) = S_e, \tag{8.4}$$

$$e = \frac{p}{\gamma - 1} + \frac{\rho v^2}{2} + \frac{B^2}{2}; \; p_t = p + \frac{B^2}{2}, \tag{8.5}$$

The conventional notation for the magnetic field **B**, the bulk velocity **v**, density ρ, pressure p, and the adiabatic index γ is used. The total energy is e, while p_t denotes the total pressure. The right–hand side of equations 8.3 and 8.4 represents the non-conservative part (usually referred to as source or sink terms) of the MHD equations. The expressions $\mathbf{S_e} = \mathbf{B} \times \eta \mathbf{j}$ and $\mathbf{S_B} = -\nabla \times \eta \mathbf{j}$ are used to introduce the effects of finite resistivity η. Here $j = \nabla \times B$ is the current density.

For simplicity, the MHD equations are cast into dimensionless form. The variables are normalized as follows: length scales by L_0, the magnetic field by B_0, the density by ρ_0, velocities by $v_0 = B_0/\sqrt{4\pi\rho_0}$, the total energy density by $e_0 = \rho_0 v_0^2/2$ and so on. For computational efficiency, the simulation is restricted to 2 dimensions, where all variables are assumed to be dependent of x and z only.

An accurate numerical solution of the non–linear MHD system requires the utilization of a shock-capturing high–order scheme for the proper calculation of the plasma dynamics. A Finite Volume Method (FVM) of Total Variance Diminishing (TVD) type, namely TVD Lax-Wendroff (TVDLF) using 2nd order in space and time is implemented in the simulator, which approximately guarantees the proper calculation of shock wave propagation and a small level of numerical oscillations induced at steep gradient regions. The TVD concept originates from the property of a linear system of hyperbolic equations in one space and one time dimension, stating that the total variance (TV) of a weak solution U (LeVeque, 2002)

$$TV(U) = \sup \sum_{j=1}^{j=N-1} |U(x_{j+1}) - U(x_j)|, \qquad (8.6)$$

does not increase with time. Here, x_j represents any set that satisfies $x_1 < x_2 < ... < x_N$ and U is defined over $[x_1, x_N]$. As suggested by Harten (1997), the discrete property of TV non-increasing

$$\sum_{j=1}^{j=N-1} |U_{j+1}^{t+1} - U_j^{t+1}| \leq \sum_{j=1}^{j=N-1} |U_{j+1}^t - U_j^t| \qquad (8.7)$$

is utilized to create a non–oscillatory FVM, in which Eq. 8.7 holds by definition. Here U_j^t represents an accurate solution U with some approximation technique at a time t and a coordinate x_j. More about TVD concept could be found elsewhere (LeVeque, 2002; Perthame and Westdickenberg, 2005); numerical tests and schemes comparison are provided in Toth (1996) and Brio and Wu (1988).

We specially take care about the magnetic field numerical divergency problem. As noted by Toth (2000), in multidimensional physical problems, the equation $\nabla \cdot \mathbf{B} = 0$ may be violated because of the spatial discretization procedure. The method utilized in this work, namely the projection scheme, removes possible $\nabla \cdot \mathbf{B}$ noise by a numerical solution of the Poisson equation $\nabla^2 \psi = \nabla \cdot \mathbf{B}^*$. The corresponding magnetic field is then updated to the divergency-free state $\mathbf{B} = \mathbf{B}^* - \nabla \psi$. The simulator performs this procedure after every third to fifth time steps of the main time advance scheme. The modifications introduced by the projection method to the conservation laws (Eqs. 8.1–8.4) are minor. A detailed description can be found in Toth (2000).

8.1.2 Benchmark problems

In order to demonstrate the reliability of the simulator, several results of benchmark simulations, performed with techniques described above, are shown in the following. This problems are commonly used for testing particular MHD codes thus supplying a sample of problems which should be handled by the numerical scheme.

Brio-Wu test

The magnetic shock tube test by Brio and Wu (1988) reveals the shock-capturing ability of the scheme (Toth, 1996). The initial condition for this test consists of an one dimensional discontinuity of all plasma parameters, which decays with time into the following waves: a fast rarefaction wave (FR) and a slow compound wave (SC), a contact discontinuity (CD), a slow shock (SS), and another FR wave. The SC wave includes a rotational discontinuity (R) and a slow shock wave. The particular initial conditions for this test are

$$\rho = 1, \mathbf{v} = 0, p = 1,$$
$$B_x = 0.75, B_z = +1, 0 < x < 0.5\,;$$
$$\rho = 0.125, \mathbf{v} = 0, p = 0.1,$$
$$B_x = 0.75, B_z = -1, 0.5 \leq x < 1\,.$$

The computation stops at $t = 0.1$. The density $\rho(x)$ for $t = 0.1$ is displayed in Fig. 8.1. Following the obtained solution, no or small spurious oscillations are generated when utilizing TVDLF. Besides, the formation of insignificant overshoots are possible (like those, displayed at the left FR and CD in Fig. 8.1). Comparing this plot with Toth (1996) shows good agreement.

Orszag–Tang test

The Orszag–Tang vortex test, also known as MHD vortex test (Orszag and Tang, 1979), presents a benchmark for the two-dimensional shock reso-

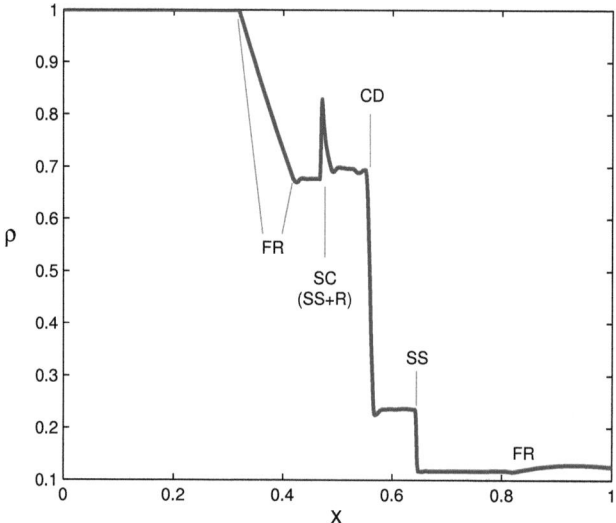

Figure 8.1: *Density $\rho(x)$ at $t = 0.1$ in Brio-Wu shock test. As expected, a fast rarefaction wave (FR), a slow compound wave (SC), a contact discontinuity (CD), a slow shock (SS), and another FR wave are found (Courtesy of A. Divin).*

lution and the $\nabla \cdot \mathbf{B} = 0$ constraint. This test originates from the study of MHD turbulence, where flows rapidly evolve from smooth initial conditions

$$\rho = \tfrac{25}{9},\ p = \tfrac{5}{3},\ v_x = -\sin z,\ v_z = \sin x,$$
$$\gamma = \tfrac{5}{3},\ B_x = -\sin z,\ B_z = \sin 2x$$

to a complicated structure of spontaneously formed shocks. The computational domain is $0 < x < 2\pi$, $0 < z < 2\pi$ with periodic boundary conditions and a 400×400 grid. A cut, corresponding to $z = 1.95$ is plotted in Fig. 8.2. The results are in good correspondence with the results from Orszag and Tang (1979) and Toth (1996).

Magnetic field diffusion test

An analytical solution exists for the evolution of a one dimensional current sheet, dividing two half–spaces with oppositely directed magnetic field and

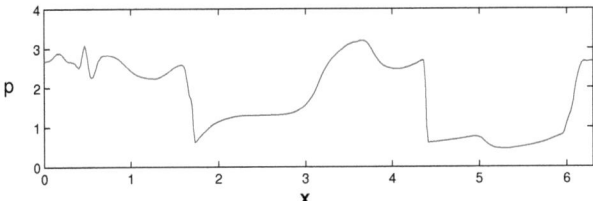

Figure 8.2: *Pressure distribution at $y = 1.95$ and $t = 3.1$ in the Orszag-Tang vortex test (Courtesy of A. Divin).*

which is affected by an uniform resistivity. We use the initial conditions

$$B_x(z) = sign(z), B_y = 0, B_z = 0,$$

at $t = 0$, then

$$B_x(z) = B_0 erf(\frac{z}{\sqrt{4\eta t}})$$

at $t > 0$. Taking $\eta = 0.01$, we start the calculation with a profile, corresponding to the analytical solution for $t = 0.5$ (thick line in Fig. 8.3) and stop at $t = 10$ (thin line in Fig. 8.3). The simulation result is shown in Fig. 8.3 with a thin dashed line. Analytical and numerical solutions differ only insignificantly.

8.2 Simulation of reconnection in plain current sheet

The following main features spontaneous reconnection in resistive MHD are commonly resolved (e.g., Ugai and Tsuda, 1977; Ugai, 1992, Abe and Hoshino, 2001), supported by theoretical considerations and recognized as an integral part of the reconnection process: acceleration of plasma up to Alfvén velocity, formation of a large-scale outflow which induce disturbances in surrounding media, as well as shock waves between the outflow region and the ambient plasma. The presented calculations display the outlined behavior well, and therefore they are in correspondence with previous works.

We took a computational domain elongated in the direction of the initial magnetic field: $L_x = 80$, $L_z = 10$ (dimensions in x and z directions, respectively). A non–uniform Cartesian grid is used, where the grid spacing varies

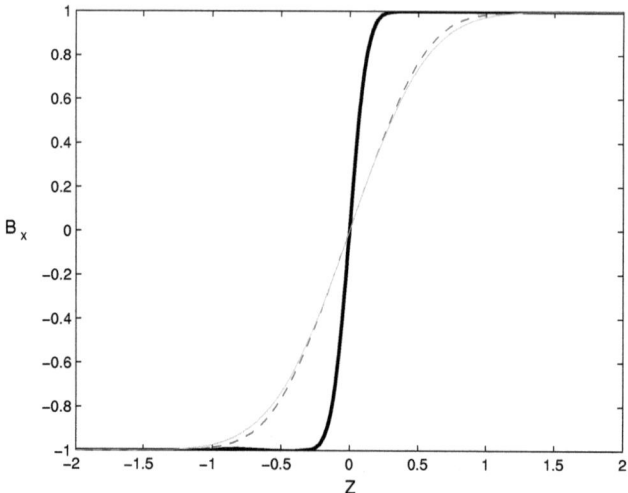

Figure 8.3: *Magnetic field diffusion test. The thick solid line shows the initial $B_x(z)$ corresponding to the analytical solution for $t = 0.5$. The thin solid line shows $B_x(z)$ for the analytical solution at $t = 10$. The thin dashed line represents $B_x(z)$ found from the simulation at $t = 10$ (Courtesy of A. Divin).*

slowly from $\Delta x = 0.04$ to $\Delta x = 0.22$ in x direction and from $\Delta z = 0.002$ to $\Delta z = 0.06$ in z direction (Fig. 8.4). The grid density is enhanced in the center of the current sheet and in the dissipative zone.

The initial equilibrium condition used in this simulation is a thin plain current sheet of Harris type

$$B_x(z) = B_0 \text{th}(z/\lambda), \quad B_0 = 1,$$

$$\rho(z) = \rho_\infty + \frac{\rho_0}{\cosh^2(z/\lambda)}, \quad \lambda = 0.06, \quad \rho_0 = \frac{2}{3}, \quad \rho_\infty = 1,$$

$$p(z) = p_\infty + \frac{p_0}{\cosh^2(z/\lambda)}, \quad \lambda = 0.06, \quad p_0 = 0.75, \quad p_\infty = 0.5.$$

The total plasma pressure $p_{tot} = p + \mathbf{B}^2/2$ is constant at $t = 0$ to support a static configuration. Boundary conditions are imposed at the domain edges by means of a ghost cell layer. At the inflow boundary ($z = \pm L_z/2$), the initial values for $\rho = 1$ and $B_x = \pm 1$ are fixed, while the other variables

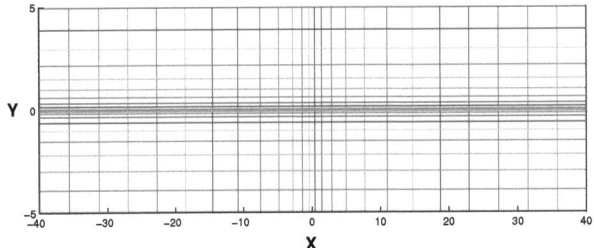

Figure 8.4: *Grid used for the reconnection simulation. The cells are shown enlarged by 20 times for a better visibility (Courtesy of A. Divin).*

are free ($\partial/\partial z = 0$), thus representing an open–flux condition. The outflow boundary ($x = \pm L_x/2$) is free, meaning that $\partial/\partial x = 0$ is applied for every variable.

In our approach, reconnection starts when the time–dependent localized resistivity is set up in the center of the current sheet

$$\eta = \eta_0 t^3 e^{-t} e^{-(x^2+z^2)/a^2} , \qquad (8.8)$$

where a is the size of the dissipative zone. During the resistivity pulse, a growth of the outflow region is observed. In Fig. 8.5, the reconnection electric field $E_R = \eta(0,0) \cdot j_y(0,0)$ is displayed.

After the localized resistivity ceased, the backfront of the outflow region starts propagating away from the former dissipative zone. The forefront of the outflow region propagates with local Alfvén velocity, whereas the velocity of the backfront is less and depends on the initial current sheet density and the reconnection pulse duration. This effect is related to the finite current sheet thickness and is not reflected in the analytical solution of the direct problem of Petschek–type reconnection.

The movement of the outflow region induces large-scale disturbances in surrounding media (Figs. 8.6 and 8.9), which are in general agreement with the theory of non-stationary MHD reconnection. Fig. 8.6 shows the x–component of the plasma velocity (intensity) and the magnetic field lines for $t = 27.8$. The plasma acceleration inside the outflow region can be clearly seen. As expected, the outflow region is bounded by slow shocks. Additionally, one can see a compression of the magnetic field lines above the outflow region, a feature which is usually referred to as a travelling compression region (TCR). In accordance with theory (Semenov et al., 1984), the disturbances

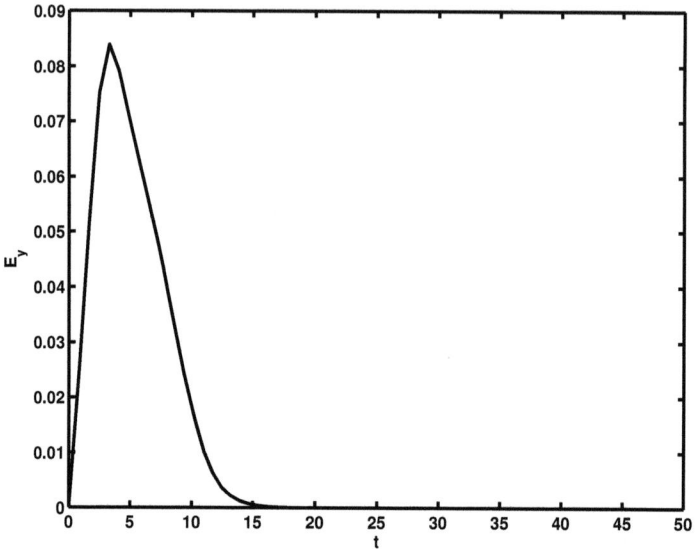

Figure 8.5: *Behavior of the reconnection electric field used for the case study (Courtesy of A. Divin).*

induced by the propagating outflow region reach the maximum in the TCR. The ambient plasma is separated from the TCR by a weak fast shock (Fig. 8.6).

In Fig. 8.9, the distribution of B_z is shown. The characteristic feature of a bipolar variation of B_z during the passage of the outflow region is seen. Referring to magnetospheric physics, similar variations are usually attributed to nightside flux transfer events (NFTEs) (Sergeev et al., 1987, 2005).

A close view of the interaction between the outflow region and the ambient plasma shows a complicated shock structure (Fig. 8.8), not resolved within the analytical theory of non–stationary reconnection. Because of the inhomogeneous background density, the propagation velocity is smaller near the center of the current sheet ($z \approx 0$) than for larger distances away from the center of the current sheet ($z > 0$). This leads to the establishment of a crab hand–like structure of the outflow region, in agreement with simulations of Ugai (1992) and Abe and Hoshino (2001). This feature is also not implemented in the analytical model. Fortunately, as can be seen in Fig. 8.8 and in plots by Abe and Hoshino (2001), this effect does not significantly alter

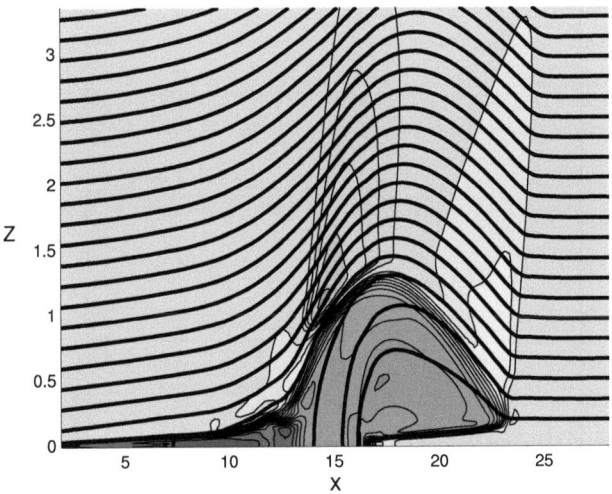

Figure 8.6: *Magnetic field line structure and v_x distribution found by the simulation. Clearly seen is the plasma acceleration in the outflow region and the generation of slow shocks (Courtesy of A. Divin).*

the outer shape of the outflow region, which is mainly responsible for the generation of the disturbances in the ambient plasma.

Since we are interested in a verification of the inverse model based on Petschek–type magnetic reconnection, we use the numerical simulation to achieve magnetic field time series corresponding to hypothetical satellite measurements. We consider three different location, namely at $x = 9.14$ and $z = 2.03$ (Case 1), $x = 14.55$ and $z = 2.03$ (Case 2), and $x = 9.14$ and $z = 3.62$ (Case 3). The selection of the locations is arbitrary, with the only restriction not to puncture the outflow region in any time during its evolution. The corresponding time series are shown in Fig. 8.9.

The expected bipolar variation is clearly visible. Additionally, the time series display the behavior expected for observations at the different location. Case 1 is closer to the reconnection site than Case 2, thus the amplitude of the B_z variation is larger for Case 2. Case 3 has the same x position as Case 1, but the z position is larger, therefore the B_z variation is smaller for Case 3, since it is further away from the current sheet. These three cases will be used

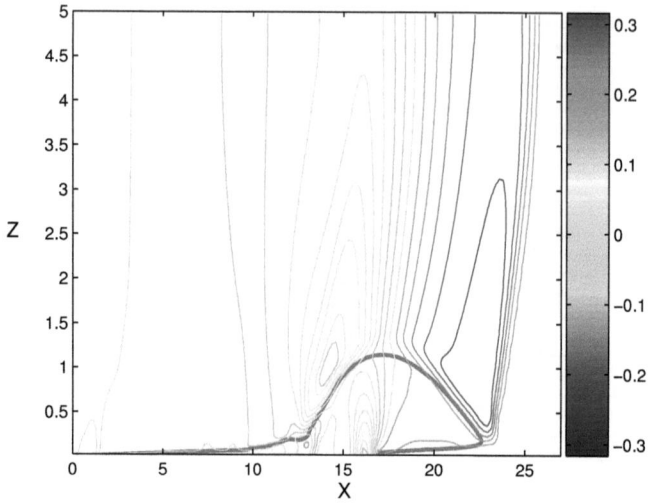

Figure 8.7: *Behavior of the B_z disturbances in the surrounding media show the characteristic bipolar variation expected for NFTEs (Courtesy of A. Divin).*

in the following as input parameters for the inverse model in order to verify it. As an additional input parameter, we use an average propagation velocity of the front of the outflow region, which is found to be about 0.75 v_A. In the analytical model, the propagation velocity is exactly v_A, but because of the finite current layer thickness, the acceleration of the plasma is less effective, resulting in a lower propagation velocity.

8.3 Reconstruction and comparison of the reconnection electric field

Using the inverse method presented in the previous section, we can now verify the method by comparing the result of the reconstruction with the numerical simulation. Therefore, we take the magnetic field time series (Fig. 8.9) as an input for Equation (5.3) and reconstruct the reconnection electric field. The result for all three cases is shown in Fig. 8.11. The black line represents the reconnection electric field used in the numerical simulation. Case 1 refers to

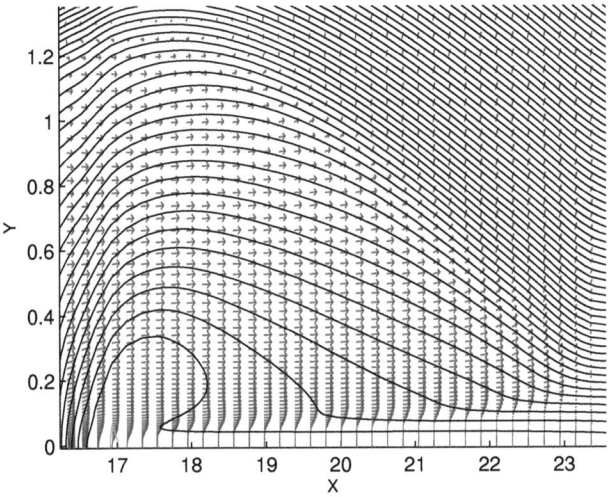

Figure 8.8: *Close-up view of the FRR region: magnetic field lines and velocity field. One can see the smaller velocity near the center of the current sheet leading to a crab-like structure of the outflow region (compare with Abe and Hoshino (2001); Courtesy of A. Divin).*

the location closest to the reconnection site, therefore the reconstruction of the reconnection electric field is the best in this case (blue line) and nearly coincides with the initial reconnection electric field. Case 2 corresponds to a large x distance from the reconnection site. In this case, the amplitude of the reconnection electric field is slightly overestimated (red line). Additionally, one can see that in this case the oscillation after the actual reconnection pulse is larger compared with the other two cases. The reason for this can be seen in Fig. 8.9. The value of the B_z component for large times is bigger than in the other cases, leading to the bigger oscillation in the result. For Case 3 it is obvious that the reconstructed electric field amplitude is slightly smaller and the time duration is slightly overestimated. This is because of the larger z distance in this case. The solution of the inverse problem for large z always tends to underestimate the amplitude and to overestimate the time duration (Semenov et al., 2005a). However, in general a good agreement between the initial reconnection electric field from the numerical simulation and the re-

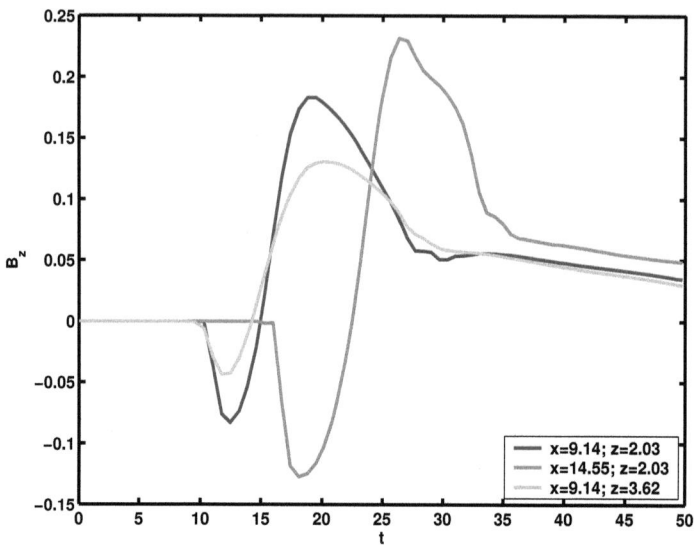

Figure 8.9: *Time series of the B_z component evaluated at different locations (Courtesy of A. Divin).*

constructed electric field from the analytical inverse model is found.

The analytical model of Petschek–type magnetic reconnection is based on several simplifying assumptions. Namely, the oppositely directed magnetic fields are separated by a tangential discontinuity, and the background quantities are assumed to be constant in the whole space. In the frame of an analytical approach, it is already complicated to deal with the simplified problem, therefore it seems not realistic to expand the analytical model to more complicated situations. In nature, the current sheet in the magnetotail has a finite thickness, and the MHD quantities are varying in the normal direction to the current sheet. Namely, B_x is zero in the center with some finite value at the edge, while the plasma density is decreasing from the center to the edge of the current sheet. In order to deal with this complicated structure, it is necessary to run numerical simulations, which are more realistic. In the numerical simulation, a finite current sheet thickness as well as an inhomogeneous distribution of the background quantities is implemented. The numerical simulation reveals different features, which are not considered in the analytical model. A main implication because of the inhomogeneous

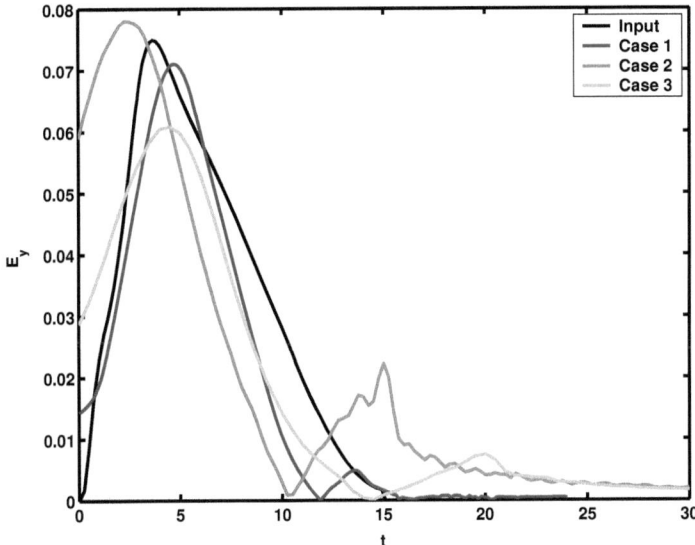

Figure 8.10: *Comparison of the reconnection electric field used in the numerical simulation (black line) and the reconstructed electric field from the magnetic field time series at three different locations (Case 1 − Case 3).*

density and velocity distribution is a distortion of the slow shocks, where a part of the outflow region is bounded by a tangential discontinuity, while at the front a crab–like structure arises, where a part of the current sheet is inside the FRR. Fig. 8.11 shows a sketch comparing the outflow region in the analytical model and the numerical simulation. However, one can see that the principle shape of the bounding structure does not change significantly, which is favorable for the analytical model.

In order to verify the inverse model to reconstruct the reconnection rate, we compared the numerical simulation and the reconstruction technique based on the inverse model. Using a predefined reconnection electric field, the numerical simulation is used to calculate time series of the magnetic field at a fixed position in space. Using these time series and the knowledge of the position, the analytical inverse model is used to reconstruct the reconnection electric field. As shown, the analytical model is able to reconstruct the main reconnection features (amplitude of the reconnection electric field, time duration of the reconnection pulse) in good agreement. This work rep-

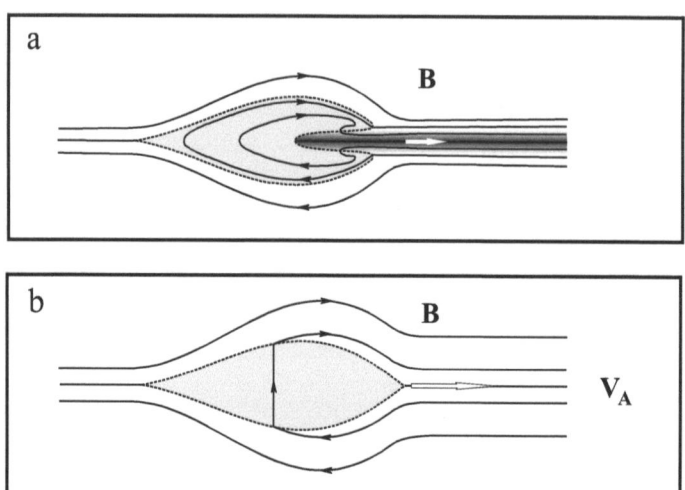

Figure 8.11: *Comparison of the configuration from the numerical simulation (a), and the analytical model (b).*

resents the first independent verification of the inverse model to reconstruct the reconnection rate.

9 Green's function of compressible Petschek–type reconnection and associated wave phenomena

The goal of this section is to study the propagation of different waves caused by reconnection in a compressible plasma in the outflow region. A detailed description of all waves occurring in the context of Petschek–type magnetic reconnection is given. According to the general Riemann problem, slow shocks and Alfvén waves will arise due to magnetic reconnection, propagating along the current sheet and causing disturbances in the ambient magnetic field. Because of the conservation of total pressure, an interaction between the upper and the lower half plane appears, leading to the development of characteristic features like fast or slow waves, tube waves, surface waves, and also side waves. In the analytical model developed by Heyn and Semenov (1996) and Semenov et al. (2004a), the disturbances in the outflow region are given as convolution integrals of a kernel with the reconnection electric field. In order to unambiguously identify the contribution of each structure, we use a delta–shaped reconnection electric field, allowing a representation of the problem in the form of Green's functions. All wave structures arising in this description can be found as poles or branch points of the according Green's function. This allows a detailed analysis of the different waves present in Petschek–type magnetic reconnection.

9.1 Wave generation in the ambient plasma environment

We perform an analysis of the case of asymmetric reconnection in a compressible plasma, using the equations for ideal compressible MHD (Eqs. 7.1 – 7.4). We follow exactly the procedure described in Section 7.1 until Eq. 7.22. But since we are now interested to study the different waves it is convenient to use a delta-function shaped reconnection electric field $E(t) = \delta(t - t_0)$ as an input function, because the convolution with a spatial extended electric field may blur different waves. By using a delta–shaped pulse, the integration over the Cagniard contour is eliminated, and only the endpoint of the integration s_{max} is needed. It can be found from solving the fourth order equation $\tau(s_{max}) = q(s_{max})z - is_{max}x$, where we choose the solution which lies in the first quadrant and satisfies the condition $\tau(s_{max}) = t$. Now the

Green's functions of the MHD variables are given as

$$\rho^{(1)} = \frac{\sqrt{\left(1+v_{Aa}^2 s_{max}^2\right)\left(1+v_{Ab}^2 s_{max}^2\right)\left(u_b^2 + c_{Sb}^2 v_{Ab}^2 s_{max}^2\right)}}{\sqrt{\left(1+c_{Sa}^2 s_{max}^2\right)\left(1+c_{Sb}^2 s_{max}^2\right)\left(u_a^2 + c_{Sa}^2 v_{Aa}^2 s_{max}^2\right)}}$$
$$\times \frac{-B_a \rho_a^{(0)} \rho_b^{(0)}}{\pi} \frac{1}{L_a + L_b} \frac{Q(s_{max})}{\tau_s}, \tag{9.1}$$

$$P^{(1)} = \frac{\sqrt{\left(1+v_{Aa}^2 s_{max}^2\right)\left(1+v_{Ab}^2 s_{max}^2\right)\left(u_a^2 + c_{Sa}^2 v_{Aa}^2 s_{max}^2\right)\left(u_b^2 + c_{Sb}^2 v_{Ab}^2 s_{max}^2\right)}}{\sqrt{\left(1+c_{Sa}^2 s_{max}^2\right)\left(1+c_{Sb}^2 s_{max}^2\right)}}$$
$$\times \frac{-B_a \rho_a^{(0)} \rho_b^{(0)}}{\pi} \frac{1}{L_a + L_b} \frac{Q(s_{max})}{\tau_s}, \tag{9.2}$$

$$B_z^{(1)} = \frac{\sqrt{\left(1+v_{Ab}^2 s_{max}^2\right)\left(u_b^2 + c_{Sb}^2 v_{Aa}^2 s_{max}^2\right)}}{\sqrt{\left(1+c_{Sb}^2 s_{max}^2\right)}} \frac{-B_a \rho_b^{(0)} i\, s_{max}}{\pi} \frac{1}{L_a + L_b} \frac{Q(s_{max})}{\tau_s}, \tag{9.3}$$

$$v_z^{(1)} = \frac{\sqrt{\left(1+v_{Ab}^2 s_{max}^2\right)\left(u_b^2 + c_{Sb}^2 v_{Ab}^2 s_{max}^2\right)}}{\sqrt{\left(1+c_{Sb}^2 s_{max}^2\right)}} \frac{-B_a \rho_b^{(0)}}{\pi} \frac{1}{L_a + L_b} \frac{Q(s_{max})}{\tau_s}, \tag{9.4}$$

where τ_s indicates a derivation of τ with respect to s, giving

$$\tau_s = \frac{\left(u^4 - v_A^2 c_S^2 + 2 v_A^2 c_S^2 u^2 s^2 + v_A^4 c_S^4 s^4\right) sz}{\sqrt{\left(1 + u^2 s^2 + v_A^2 c_S^2 s^4\right)\left(u^2 + v_A^2 c_S^2 s^2\right)^3}} - ix. \tag{9.5}$$

Using the exact compressible wave motions driven by sources created by magnetic reconnection, which are moving along the x–axis, the system response function is derived in a way that it can be used for linear plane wave problems. This behavior will be investigated in the following in more detail.

The poles and branch points found in the Green's function arise from four different terms in Eqs. (9.1)–(9.4):

- from the source function $Q(s_{max})$, describing Alfvén waves and slow shocks,
- from $L_a + L_b$ leading to the appearance of surface waves,
- from τ_s giving a fast body wave, and
- from the branch points of the first term in Eqs. (9.1)–(9.4) leading to waves travelling with fast, slow, and tube speed.

Tube waves are similar to sound waves, except that they expand in regions of increased gas pressure. This lateral expansion reduces the propagation speed of the tube wave below the sound speed c_S (Defouw, 1976). The tube speed is given as

$$v_T = \frac{c_S^2 v_A^2}{c_S^2 + v_A^2}. \tag{9.6}$$

It might be noticed that $L_a + L_b$ is the dispersion relation of the Kelvin–Helmholtz instability. This instability arises if a large shear flow destabilizes the surface. However, in order to study magnetic reconnection, we suppose Kelvin–Helmholtz stability, because otherwise reconnection is likely to be suppressed.

9.2 The case of symmetric magnetic fields

In order to investigate the structure of the excited waves and shocks, we start with a simple configuration and proceed to more complicated configurations in the following. As a first example, symmetric, antiparallel magnetic fields with $B_a = 1$ and $B_b = -1$ is used for a low plasma β of $\beta_a = 0.1$ and $\beta_b = 0.1$. In Fig. 9.1, the density distribution at $z = 0$ is shown along the current sheet (x-axis) for $t = 4$. We can see different wave and shock structures, arising from poles and branch points in Eq. (9.1). Starting from $x = 0$ in Fig. 9.1, we first find the contribution from the branch points of the first fraction in Eq. (9.1), where the branch point of the first expression is called BP1, and so on. At $x = 1.1$, the tube wave of BP3 is found (TW$_a$, TW$_a$), while BP1 and BP2 coincide for symmetric conditions giving the slow waves from the upper and the lower half plane at $x = 1.15$ (SW$_a$, SW$_b$; see Table 9.1). For the symmetric case, the slow shocks and Alfvén discontinuities from both half planes (S$_a$, S$_b$, A$_a$, A$_b$) coincide at $x = 4$. It can be seen that the tube speed (v_T) is smaller than both the sound (c_S) and the Alfvén speed (v_A).

From Eq. (9.5) one can see that for $z > 0$ a fast shock arises, giving the first front seen in Fig. 9.2. Additionally, all other wave and shock structures identified for $z = 0$ can be found in Fig. 9.2. Additionally, it can be seen that for small z at the first front a compression appears (red color). This is because of the propagation of the extended flux tube through the plasma, where the media gets compressed at the front and the plasma becomes more dense and gets a higher temperature. But for large z, rarefaction takes place (blue color), because after the flux tube passed by and the flux was reconnected, the magnetic field intensity decreases below the initial value, leading to a rarefaction with an associated decrease of the plasma density and temperature.

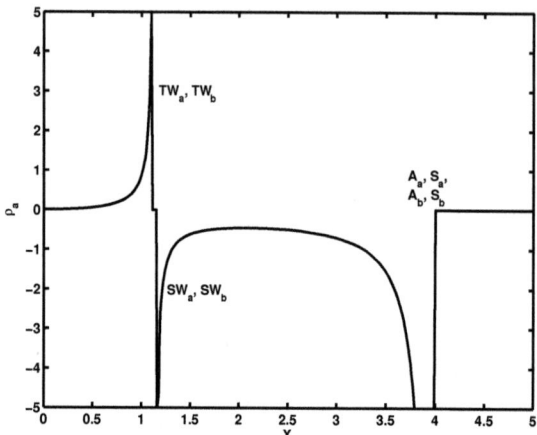

Figure 9.1: *Density distribution at $z = 0$. Seen are different wave and shock structures, namely the tube wave from branch point 3 (TW_a, TW_b), the slow waves from the upper and the lower half plane (SW_a, SW_b), as well as the slow shocks and Alfvén discontinuities from both half planes (S_a, S_b, A_a, A_b).*

Wave	Velocity	Location
A_a	1	4
S_a	1	4
A_b	1	4
S_b	1	4
SW_a	0.289	1.156
SW_b	0.289	1.156
TW_a	0.275	1.1
TW_b	0.275	1.1

Table 9.1: *Location and velocity of the wave and shock structures found for the case of symmetric, antiparallel magnetic fields (compare with Fig. 9.1).*

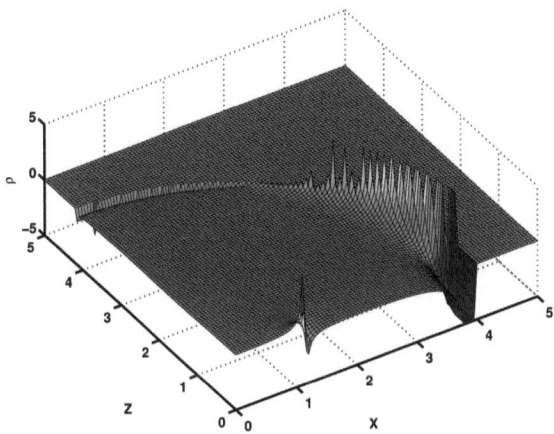

Figure 9.2: *Density distribution in the x–z–plane. The first front is a fast wave arising from τ_s. Also clearly visible are contribution from the Alfvén discontinuities, the slow shocks, the slow waves, and from the tube waves (which all coincide for the upper and the lower half plane in the symmetric case).*

In this special case, the solution for the lower half plane is completely symmetric.

9.3 The case of asymmetric magnetic fields

9.3.1 Low plasma β case

In this section, we consider antiparallel magnetic fields with different field strength in the upper and the lower half plane. As a first example, the case of a low plasma β is considered. We use $B_a = 1$, $B_b = -0.5$, $\beta_a = 0.1$, and $\beta_b = 3.4$. Due to the slightly asymmetric conditions, slow shocks will appear and no rarefaction waves.

The source function $Q(s_{max})$ gives poles corresponding to the propagation of the Alfvén discontinuity and the slow shock in the upper and the lower half space. The phase velocity of these structures is shown in Table 9.2. The structure with the fastest propagation velocity is the Alfvén discontinuity ($w_{Aa} = 1$) followed by the slow shock ($w_{Sa} = 0.894$) in the upper half space. These contributions are seen in Fig. 9.3 at $x = 4$ and $x = 3.576$, respectively. The contributions of the Alfvén wave and the slow shock generated in the

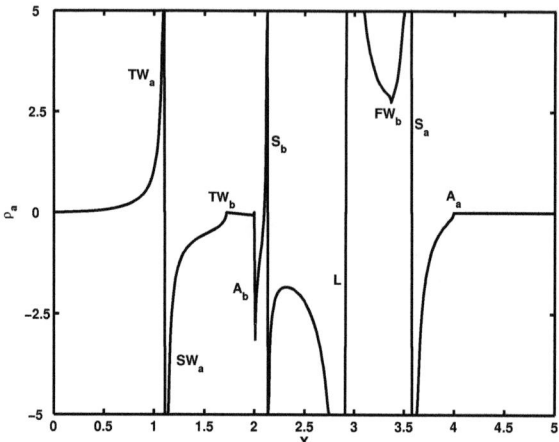

Figure 9.3: *Density distribution at $z = 0$ for asymmetric magnetic fields and low plasma β in the upper half plane. Seen are different wave and shock structures, namely the tube waves from both half planes (TW_a, TW_b), the slow wave from the upper and the fast wave from the lower half plane (SW_a, FW_b), the slow shocks and Alfvén discontinuity from both half planes (S_a, S_b, A_a, A_b), as well as the surface wave from $L_a + L_b$ (L).*

Name	Velocity	Location
A_a	1	4
S_a	0.894	3.576
A_b	0.5	2
S_b	0.533	2.132
SW_a	0.289	1.15
FW_b	0.838	3.35
TW_a	0.275	1.1
TW_b	0.425	1.7
L	0.728	2.911

Table 9.2: *The different wave structures for the low β case with corresponding propagation velocities and the location at $t = 4$ for the upper half plane (compare with Fig. 9.3).*

lower half space, which are also visible in the upper half space, have phase velocities of $w_{Ab} = 0.5$ and $w_{Sb} = 0.533$, respectively, and can be found at $x = 2$ and $x = 2.132$ in Fig. 9.3. A surface wave comes from L_a+L_b, since now there is a shear velocity between the upper and the lower half plane. It has a propagation velocity of 0.728 and is seen at $x = 2.911$ in Fig. 9.3. From the fraction in Eq. 9.1, three branch points occur. The are named in the same way as in the previous section. The phase velocities are $c_{Sa} = 0.288$, $c_{Sb} = 0.838$, and $v_{Ta} = 0.275$ for BP1, BP2, and BP3, respectively. This corresponds to $x = 1.15$, $x = 3.35$, and $x = 1.1$ in Fig. 9.3. Similar to the symmetric case, the contribution of BP1 is the slow wave for the upper half plane, since for $\beta < 1$, the slow wave velocity corresponds to the sound speed. In contrast, BP2 gives a fast wave in the lower half plane, since $\beta > 1$ in the lower half plane, so that the sound speed corresponds to the fast wave speed in this case (compare with Table 9.2). The tube wave is again the slowest wave in both half planes. Now all seven wave structures appearing in this configuration are specified. For the lower half plane, the wave structure is similar, only the contribution from the tube wave differs, leading to a phase velocity of $v_{Tb} = 0.425$, corresponding to a location at $x = 1.7$ in Fig. 9.3.

The distribution of the variations for the plasma density, the total pressure, the B_z-component, and the v_z-component for the upper half plane are shown in Fig. 9.4. For the plasma density, one can see all disturbances from Fig. 9.3 and the fast front from τ_s. The expression for the pressure (Eq. 9.2) shows a similar structure as Eq. 9.1, but only branch points BP1 and BP2 are occurring. Therefore, no disturbances from a tube wave (corresponding to BP3) can be found in the pressure distribution. For the disturbances of the z-component of the magnetic field (9.3) and the z-component of the velocity (9.4), only BP2 appears. For increasing z, most of the fronts disappear. Only the surface wave from L_a+L_b and the front of the fast body wave give significant contributions for $z > 0.2$, which can be also seen in Fig. 9.4. In Fig. 9.5, the structure of the boundary layer according to the solution of the Riemann problem is shown. The structure of the discontinuities from right to left is $A_a S_a C S_b A_b$, where the positions of these features corresponds to the locations given in Table 9.2. The position of the contact discontinuity (C) is correlated with the surface wave caused by L_a+L_b.

9.3.2 High plasma β case

For comparison, we consider now the case of a high plasma β and slightly asymmetric magnetic fields. We use $B_a = 1$, $B_b = -0.35$, $\beta_a = 5.0$, and $\beta_b = 47.98$. The behavior of the plasma density at $z = 0$ for both half spaces is shown in Fig. 9.6. In this case, a different behavior compared with the

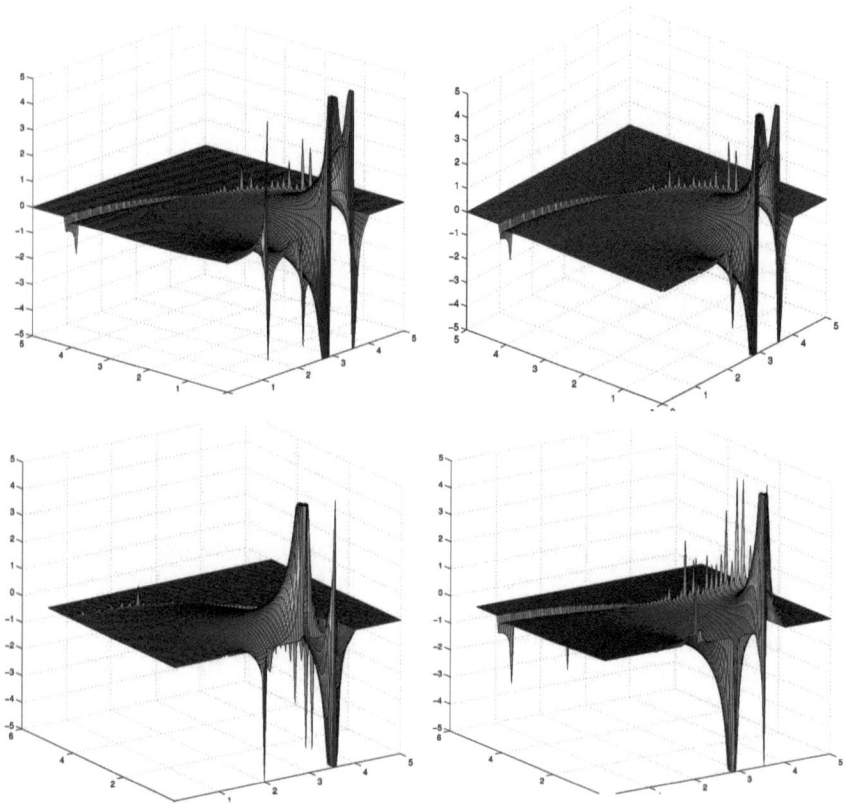

Figure 9.4: *Variations of the plasma density (upper left panel), the total pressure (upper right panel), the z-component of the magnetic field (lower left panel), and the z-component of the velocity (lower right panel) for the low plasma β case.*

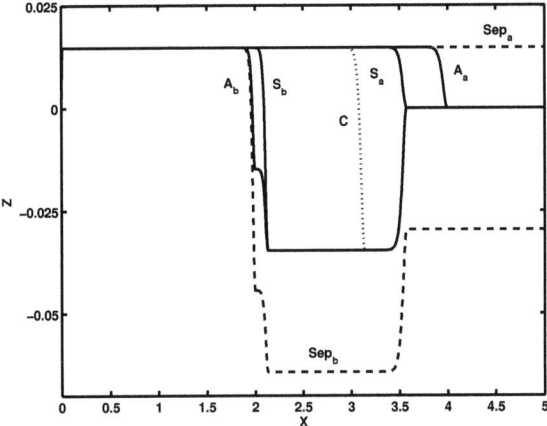

Figure 9.5: *Shock structure according to the solution of the Riemann problem. The dashed lines indicate the location of the seperatices (Sep_a, Sep_b), while the solid lines show the Alfvén waves (A_a, A_b), and the slow shocks (S_a, S_b), while the dotted line indicates the contact discontinuity (C).*

low plasma β case is found (Fig. 9.6 and Table 9.3). The Alfvén wave again propagates with a phase velocity of $w_{Aa} = 1$, followed by the slow shock, which has only a slightly smaller propagation speed, namely $w_{Sa} = 0.978$. Thus, at $t = 4$, the location of the fronts are still close to each other at $x = 4$ and $x = 3.912$, respectively. For the lower half space, the phase velocities of the Alfvén wave and the slow shock are nearly the same ($w_{Ab} = 0.35$ and $w_{Sb} = 0.354$). Therefore, the perturbations caused in the upper half space are still merged at $t = 4$ and are located at $x = 1.4$ and $x = 1.416$. The surface wave from the pole in $L_a + L_b$ is again located between the poles of the source term from the upper and the lower half space at 2.911, corresponding to a phase velocity of 0.728. The phase velocity of the sources from BP1 and BP2 is $c_{Sa} = 2.04$ and $c_{Sb} = 2.212$, respectively, so that these disturbances are located at $x = 8.16$ and $x = 8.85$ at $t = 4$. This means that there are perturbations excited by sources moving with superalfvénic speed, since in the case of $\beta > 1$ the fast wave speed corresponds to the sound speed (Fig. 9.6 for $x > 4$). BP3 leads to a tube wave moving with a phase velocity of $v_{Ta} = 0.898$, which is located at $x = 3.59$ in Fig. 9.6. For the lower half plane, only the phase velocity of the tube wave shifts to $v_{Tb} = 0.345$ and is only slightly smaller than the Alfvén wave in the lower half plane.

The distribution of the variations for the plasma density in the upper half

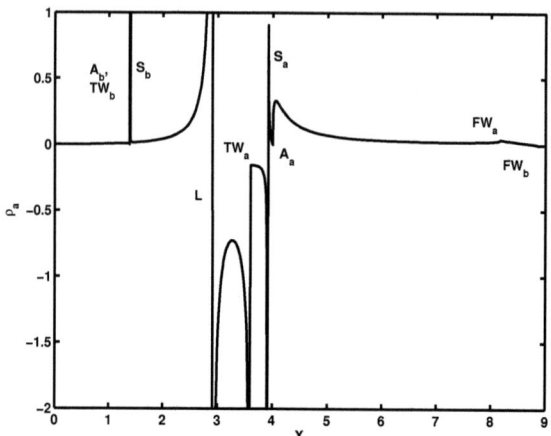

Figure 9.6: *Density variations in the upper half space for $z = 0$ in the high β case. Clearly visible is the fact that for a high plasma β the contribution from BP1 and BP2 give a fast wave (FW_a, FW_b). It is followed by the other perturbations as listed in Table 9.3.*

Wave	Velocity	Location
A_a	1	4
S_a	0.978	3.912
S_b	0.354	1.416
A_b	0.35	1.4
L	0.728	2.911
FW_a	2.04	8.16
FW_b	2.212	8.85
TW_a	0.898	3.59
TW_b	0.345	1.38

Table 9.3: *The different wave structures for the high β case with corresponding propagation velocities and the location at $t = 4$ for the upper half plane (compare with Fig. 9.6).*

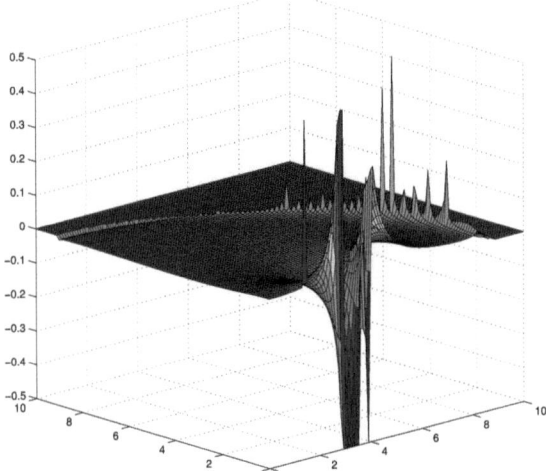

Figure 9.7: *Density distribution in the x–z–plane. The first front is a fast wave arising from τ_s and from the BP2. Also clearly visible are the contributions from the Alfvén discontinuities, the slow shocks, the surface and the tube wave.*

plane is shown in Fig. 9.7. In this case, the first wave front is due to the propagation of the disturbance caused by BP1 and BP2. The other observed features in this figure can be explained similar as in the low β case.

9.4 The appearance of side waves

If some waves in the lower half plane are propagating with a velocity faster than the fastest one in the upper half plane, so–called side waves arise in the upper half plane. They are caused by the disturbances of the propagating waves in the lower half plane and show the characteristic form of a Mach cone in the upper half plane. In order to achieve a fast propagation velocity in the lower half plane, we increase the plasma density in the upper half plane $\rho_a = 3.0$, while $\rho_b = 0.5$. The magnetic fields are chosen to be symmetric as $B_a = 1.0$ and $B_b = -1.0$, while $\beta_a = 0.5$ and $\beta_b = 0.5$.

For these parameters, it is obvious that the disturbances in the lower half plane are propagating faster. The corresponding velocities for both half planes are shown in Table 9.4 and Fig. 9.8. The fastest wave in the lower half plane is the Alfvén wave ($w_{Ab} = 1.414$), followed by the slow shock

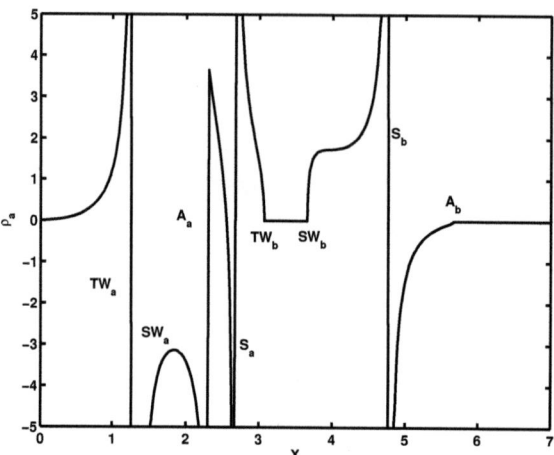

Figure 9.8: *Density variations in the upper half space for $z = 0$ in the case where side waves arise.*

Wave	Velocity	Location
A_a	0.577	2.31
S_a	0.668	2.67
S_b	1.192	4.77
A_b	1.414	5.66
SW_a	0.373	1.49
SW_b	0.913	3.65
TW_a	0.313	1.25
TW_b	0.767	3.07

Table 9.4: *Different wave structures with corresponding propagation velocities and the location at $t = 4$ for the upper half plane (compare with Fig. 9.8) for the side wave case.*

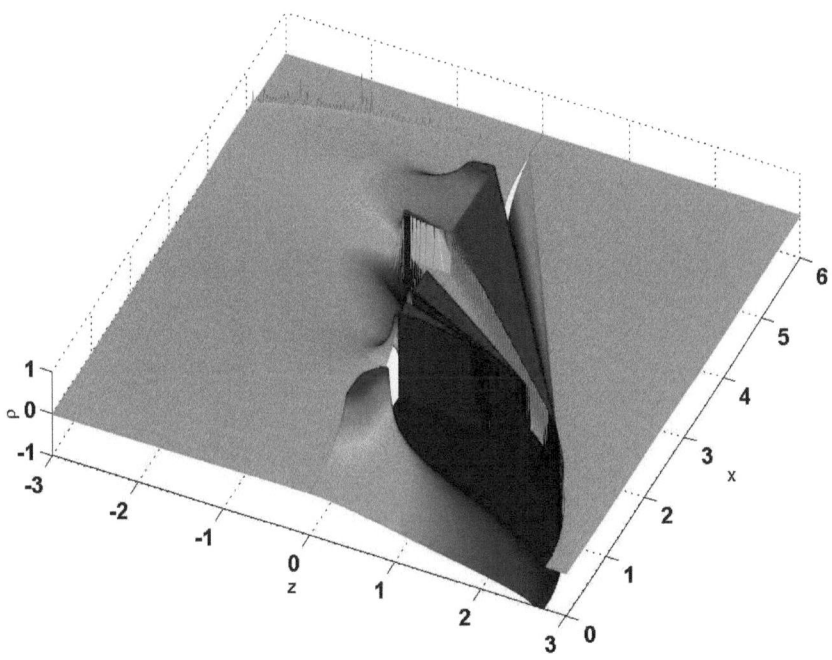

Figure 9.9: *Density distribution in the x–z–plane. The first side wave arises from the Alfvén discontinuity, the second from the slow shock, the third one from the slow wave. The fourth side wave, which comes from the tube wave is separated from the other side waves by a lagoon. After the four side waves, the wave structure is similar as in the previous examples.*

($w_{Sb} = 1.192$), the slow wave ($c_{Sb} = 0.913$), and the tube wave ($v_{Tb} = 0.767$). Compared with the fastest disturbance in the upper half plane, the slow shock ($w_{Sa} = 0.668$), all disturbances in the lower half plane are moving faster. This leads to the occurrence of four side waves in the upper half plane, caused by the four structures from the lower half plane. These side waves form a Mach cone with an apex angle proportional to the fast wave speed in the upper half plane, which can be seen in Fig. 9.9. Additionally, between the side wave caused by the slow wave and the side wave from the tube wave, a lagoon can be observed, where the values for the density disturbances are exactly zero. Since we consider symmetric magnetic fields, no contribution from $L_a + L_b$ can be observed.

10 Conclusions and Outlook

In this work, a method to reconstruct the reconnection rate is developed for incompressible and compressible plasma. Application of the method to various Cluster measurements in the magnetotail as well as near the cusp is shown. Additionally, a verification of the method by using a numerical MHD simulation of plasmoids in the magnetotail was performed. Also the wave generation by Petschek–type magnetic reconnection is discussed.

Based on the analytical model developed by Heyn and Semenov (1996) and Semenov et al. (2004a) a direct model for incompressible plasma is developed. It is used to achieve an expression for the displacement vector in coordinate–time space taking advantage from the so–called Cagniard–deHoop method. An expression for the displacement vector is found in form of a convolution integral. This is a favorable form for the treatment of the problem in the frame of an inverse model. We use Tikhonov regularization and introduce a regularization parameter. This technique is used to reconstruct the reconnection electric field out of magnetic field measurements of satellites. In order to apply the theoretical model to measured data, it is necessary to know crucial quantities like the distance between the satellite and the reconnection site, or the Alfvén velocity. However, the application of an analytical model requires to make some simplifications, which should be mentioned in the following.

The x–distance between the satellite and the reconnection site is determined by using a minimization routine. There exists the possibility that more that one minimum occurs, meaning that there is no single solution for the problem considered. In this case, the routine may give a wrong result. To avoid this problem, we applied our method only to the range of x–distances, where reconnection most likely takes place, namely the NENL to a distance less than 35 R_e, and run it with different starting points. It should be noted that a small difference in the x–distance does not influence the shape of the disturbances significantly. Additionally, we assume that the perturbations in the magnetic field are moving approximately with Alfvén velocity based on the assumption of a homogeneous background density in the magnetotail. If the density changes significantly between the point of observation and the starting point of the disturbances, the estimated Alfvén velocity may differ from the real one. Since the Alfvén velocity is used for the normalization of the length scales, a variation of the Alfvén velocity will also give a variation of the spatial distances. If the Alfvén velocity decreases, also the length scales will decrease. Another simplification is the assumption of the incompressibility of the plasma. Since this is a rough approximation for space plasmas, we extend the model to a compressible plasma. As already predicted (Semenov et al., 2005a), the reconnection rate is smaller than inferred from the incompressible model. Also the distance between the reconnection site and the satellite decreased, while the duration of the pulses increase. All these features are consistent with the qualitative estimates done by Semenov et al. (2005a). However, a significant reduction of the noise resulting from the solution of the inverse problem is achieved compared with the incompressible model. Also the variations of the reconstructed reconnection rate and site between the different satellites decreased.

Application to Cluster measurements were done for several cases. Most extensively investigated is a series of NFTEs on September 8th, 2002. The incompressible model gives an amplitude of the reconnection electric field in the range of 1–2 mV/m. The time duration of the reconnection pulse is in the order of 20–30 s, while the reconnection site is located at about 29 R_e tailwards. The compressible model gives slightly different results according to the discussion above. The amplitude of the reconnection electric field is in the range of 0.9–1.2 mV/m. This amplitude of the reconnection electric field is consistent with estimations of the magnetotail reconnection rate obtained from ground–based measurements. The time duration of the reconnection pulse is in the order of 50 s, while the reconnection site is located at about 25 R_e tailwards. Analysis of NFTEs on September 26th, 2005 is also in good agreement with theoretical considerations. In this case, the reconnection is located close to the satellites. However, since the z–distance between the spacecraft and the current sheet is poorly known, which influences the amplitude of the reconnection electric field. The reconstruction reveals that the time duration of the event is about 2 min, and the reconnection process was initiated slightly before 8:42:30 UT at a location somewhere around 14 R_e. Also the results for high–latitude FTEs give reasonable results.

From the reconstruction of FTEs/NFTEs, no verification of the model can be achieved. Thus, a numerical MHD simulation of plasmoid propagation is used. In order to verify the inverse model to reconstruct the reconnection rate, we compared the numerical simulation and the reconstruction technique based on the inverse model. Using a predefined reconnection electric field, the numerical simulation is used to calculate time series of the magnetic field at a fixed position in space. Using these time series and the knowledge of the position, the analytical inverse model is used to reconstruct the reconnection electric field. As shown, the analytical model is able to reconstruct the main reconnection features (amplitude of the reconnection electric field, time duration of the reconnection pulse) in good agreement. This work represents the first independent verification of the inverse model to reconstruct the reconnection rate.

Additionally, a method to analyze the behavior of waves caused by magnetic reconnection in a compressible plasma is presented. It is shown that there appear different wave and shock structures, namely Alfvén waves and slow shocks in the upper and the lower half plane caused by the source term, surface waves caused by L_a+L_b, the fast body wave from τ_s, as well as perturbations due to branch points giving structures moving with sound and tube speed. For the case of symmetric antiparallel magnetic fields a relatively simple structure is found: the Alfvén and slow shocks from both sides are merged and establish together with the fast wave from τ_s the first wave front. This front is followed by a merged slow shock from the contributions of both sides. As predicted, the slowest wave is a tube wave, which is propagating with a slightly smaller velocity than the slow shocks. The contribution from τ_s gives a fast body wave. Because of the symmetric conditions, no surface wave from L_a+L_b arises. For the case of asymmetric antiparallel magnetic fields, a more complex behavior is found. In this case, the Alfvén and slow shocks are not merged anymore, so that each of them causes a perturbation in the ambient plasma. In this case a surface wave appears because of the contribution from L_a+L_b. Depending on the value of β, the contribution from the branch points gives a slow wave ($\beta < 1$) or a fast wave ($\beta > 1$), while the third branch point gives a tube wave, which propagates always slower than the Alfvén and the slow wave. Again, τ_s leads to a fast body wave. If the wave and shock structures in the lower half plane are propagating faster compared with the upper half plane, side waves arise. Under certain circumstances, also a lagoon–type solution can appear.

This work presents a first attempt to reconstruct the reconnection electric field from Cluster measurements. All results are in good agreement with theoretical considerations, with observations, and with numerical simulations.

11 Appendix: The general Riemann problem

The Riemann problem we are analyzing here, concerns the decay of a tangential discontinuity into a system of MHD shocks and waves. If we introduce the eight–dimensional MHD state vector $U(\rho, p, \vec{v}, \vec{B})$, which contains information about all eight variables in the MHD system of equations (4.7–3.19), then we can replace the set of equations with one single equation of matrix form (Akhiezer et al., 1975)

$$\frac{\partial}{\partial t}U + M\frac{\partial}{\partial x}U = 0, \qquad (A.1)$$

where M is a symmetric matrix whose elements are functions of the variables mentioned above. This makes clear, that the set of ideal MHD equations forms a quasi–linear, symmetric hyperbolic system. In the MHD description of a plasma, various wave modes can be excited. These are the fast and slow waves, Alfvén waves, and two non–propagating structures: the tangential pressure balance and the entropy wave (Barnes, 1983). A characteristic feature, that arises in the non–linear theory of simple MHD waves is the formation of shocks and discontinuities, resulting from the combination of steepening and dispersion of the wave modes. Thus, the compression waves of the fast and slow modes steepen into fast and slow shocks under certain conditions. The Alfvén, tangential and entropy wave structures do not steepen in the same way, but if they are considered sufficiently thin, they are classified as discontinuities: the rotational, tangential and contact discontinuities, respectively.

Following Akhiezer et al. (1975) we can see, if MHD waves are formed in a single source not more than three waves can move in the same direction: in front a fast wave, behind it follows an Alfvén discontinuity and, finally, behind them, there moves a slow wave. Of course, some of the waves enumerated here may not be present.

We can consider a state vector $U(x,t) = U(\rho(x,t), p(x,t), \vec{v}(x,t), \vec{B}(x,t))$. Proposing that at $t = 0$ the state vector jumps from one constant value to another constant value across $x = 0$, the problem describing the subsequent behaviour of this discontinuity is often called the generalized Riemann problem in MHD. Since here the state vector suffers a discontinuity, it is necessary to analyse the following problem by using the Rankine–Hugoniot conditions. As our initial state we consider a tangential discontinuity for which the normal components of the magnetic field and plasma velocity are zero ($B_n = 0$, $v_n = 0$). Across such a discontinuity the following jump conditions hold (Heyn et al., 1988):

$$\left[\!\left[B_t \right]\!\right] \neq 0, \tag{A.2}$$

$$\left[\!\left[v_t \right]\!\right] \neq 0, \tag{A.3}$$

$$\left[\!\left[p \right]\!\right] \neq 0, \tag{A.4}$$

$$\left[\!\left[p + \frac{B_t^2}{8\pi} \right]\!\right] = 0. \tag{A.5}$$

These equations express the fact, that the density and the tangential components of the magnetic field and velocity may change arbitrarily, subject only to the requirement, that the total pressure stays constant. Note that there is no mass flow and no magnetic connection across the tangential discontinuity, there will be no electric field component along it ($E_t = 0$).

Now we assume, that a tangential electric field E^* is introduced in a certain region, equivalent to the switching on of reconnection (Biernat et al., 1987), in other words, there appears a normal component of the magnetic field B_n. This B_n will propagate away from the initial reconnection site, and as a consequence, the tangential discontinuity decays into a system of MHD wave modes. This follows from the fact, that a discontinuity characterized by the violation of conservation laws is unstable and it decays into a system of other discontinuities and shocks. The most general scheme of a decay looks like $S^+(R^+)AS^-(R^-)C(T)S^-(R^-)AS^+(R^+)$ (Akhiezer et al., 1975), where S^+, S^- are fast and slow shocks, R^+, R^- are fast and slow rarefaction waves, and A, C, T are Alfvén, contact and tangential discontinuities, respectively.

As shown in Figure 11.1, a suitably chosen combination of these will enable the transition from one external region to the other modelled in the presence of normal field and flow components. The diffusion region, in which ideal MHD is not valid, is assumed to be sufficiently small so that the ideal MHD waves can be considered as emanating from a singular line, the reconnection line. The outer edge of the layer is bounded by an Alfvén wave, then follows a slow shock or a slow expansion wave and finally a contact discontinuity, which separates the two external regions.

The solution of the Riemannian problem is defined by the vector (Heyn and Semenov, 1996; Semenov et al., 2004a, and references therein)

$$\mathbf{h} \equiv \text{sgn}(mB_n)(\tilde{\mathbf{v}}_0 - \mathbf{v}_0) + \tilde{\mathbf{v}}_{A0} + \mathbf{v}_{A0}, \tag{A.6}$$

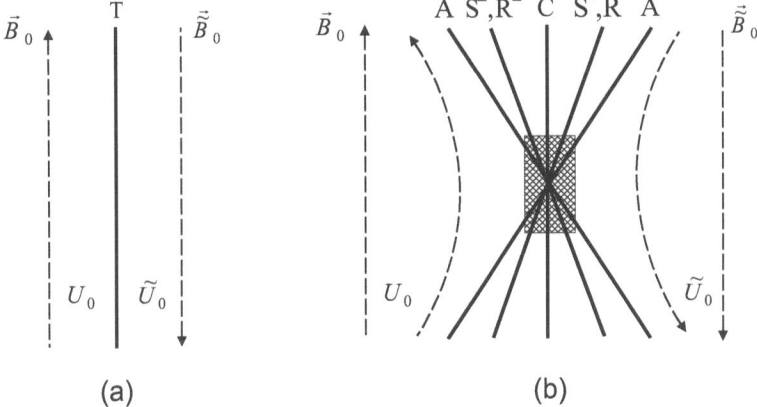

Figure 11.1: *A tangential discontinuity (a) seperating two uniform but different plasma and field regions. (b) The structure of the reconnection layer after introducing a normal magnetic field component. The diffusion region is the hatched area in (b).*

where $m \equiv \rho(v_n - D)$ is the mass flux through the discontinuity and D is the speed of the discontinuity. The magnetic field $\mathbf{B}_1$ downstream of the Alfvén discontinuities is parallel to the vector $\mathbf{h}$, i.e.,

$$\mathbf{b}_1 = \frac{\mathbf{h}}{\|\mathbf{h}\|}. \tag{A.7}$$

The MHD quantities in region AS are

$$\mathbf{B}_1 = \mathbf{b}_1 B_0, \tag{A.8}$$

$$\mathbf{v}_1 = \mathbf{v}_0 + \operatorname{sgn}(mB_n)\left(\mathbf{b}_1 v_{A0} - \mathbf{v}_{A0}\right), \tag{A.9}$$

$$\rho_1 = \rho_0, \tag{A.10}$$

$$p_1 = p_0, \tag{A.11}$$

$$\left(\frac{m}{B_n}\right)_A = \pm\sqrt{\frac{\rho}{4\pi}}, \tag{A.12}$$

$$\mathbf{w}_A \equiv \mathbf{v} - \frac{1}{\rho}\left(\frac{m}{B_n}\right)_A \mathbf{B}. \tag{A.13}$$

Here, $\mathbf{w}_A$ is the de Hoffmann-Teller velocity of the Alfvén discontinuity. Similar formulas are valid for the region $\tilde{A}\tilde{S}$.

127

To find the MHD quantities in the region $SC\tilde{S}$ it is convenient to use as a key parameter the change in magnetic strength across the slow shock $\eta = B_2/B_0$, and $\tilde{\eta} = \tilde{B}_2/\tilde{B}_0$. Since the magnetic field is continuous across the contact discontinuity, one can write $\eta B_0 = \tilde{\eta}\tilde{B}_0$. Then, the parameter η can be calculated from

$$\|\mathbf{h}\| = v_{A0} G(\eta) + \tilde{v}_{A0} G(\tilde{\eta}) \tag{A.14}$$

where the function G is defined as

$$G(\eta) = 1 - \sqrt{(1-\eta)\left(1 - \left[\frac{1-\eta^2}{2\beta + (\gamma-1)(1-\eta)}\right]\eta\right)}, \tag{A.15}$$

with $\beta = c_s^2/v_{A0}^2$ and γ the polytropic index. Note that across the slow shock the magnetic strength decreases, hence $\eta < 1$ and $\tilde{\eta} < 1$. After η had been found, the MHD values downstream of the slow shock are

$$\mathbf{B}_2 = \mathbf{b}_1 \eta B_0, \tag{A.16}$$

$$\mathbf{v}_2 = \mathbf{v}_0 - \text{sgn}(mB_n)\left[\mathbf{v}_{A0} - \mathbf{b}_1 v_{A0} G(\eta)\right], \tag{A.17}$$

$$\frac{\rho_2}{\rho_0} = 1 + \frac{1-\eta^2}{2\beta + (\gamma-1)(1-\eta)}, \tag{A.18}$$

$$p_2 = p_0 + \frac{B_0^2}{8\pi}\left(1 - \eta^2\right), \tag{A.19}$$

$$\left(\frac{m}{B_n}\right)_S = \pm\frac{1}{\sqrt{4\pi}}\sqrt{\frac{\eta-1}{\frac{\eta}{\rho_2} - \frac{1}{\rho_1}}}, \tag{A.20}$$

$$\mathbf{w}_S \equiv \mathbf{v} - \frac{1}{\rho}\left(\frac{m}{B_n}\right)_S \mathbf{B}. \tag{A.21}$$

The local horizontal elongation of the shocks will be of first order in B_n/B_0, say,

$$z = f(t, x, y), \tag{A.22}$$

and, therefore, normal vector $\mathbf{n}$ and shock speed D are

$$\mathbf{n} = (-f_x, -f_y, 1), \quad D = f_t, \tag{A.23}$$

since $\sqrt{1 + f_x^2 + f_y^2} \approx 1$ in lowest order. Therefore, mass flux and normal magnetic field are

$$m = \rho(v_n - D) = \rho\left(-\mathbf{v}^0 \cdot \nabla f + \frac{\partial \zeta}{\partial t} + \mathbf{v}^0 \cdot \nabla \zeta - \frac{\partial f}{\partial t}\right), \tag{A.24}$$

$$B_n = (-\mathbf{B}^0 \cdot \nabla f + \mathbf{B}^0 \cdot \nabla \zeta). \tag{A.25}$$

Then the Walén relation at the Alfvén discontinuity (A.13) and at the slow shock (A.21) lead to the equation

$$\left(\frac{\partial}{\partial t} + \mathbf{w}^{(0)} \cdot \nabla\right)(f - \zeta) = 0, \tag{A.26}$$

with a solution

$$f(t, x, y) = \zeta(t, x, y) + \Phi\left(t - \frac{x}{w_x^{(0)}}, y - \frac{w_y^{(0)}}{w_x^{(0)}} x\right), \tag{A.27}$$

where the velocity **w** had been defined above. The y-component of the electric field within any discontinuity in its rest frame is

$$\begin{aligned}
E^*(t, x, y) &:= \frac{1}{c}((\mathbf{v} - D\mathbf{n}) \times \mathbf{B}) \cdot (\mathbf{n} \times \mathbf{e}_x) \\
&= \frac{B_x^{(0)}}{c}(v_n - D) - \frac{B_n}{c}(v_x^{(0)} - Dn_x) \\
&\approx \frac{B_x^{(0)}}{c}(v_n - D) - \frac{B_n}{c}v_x^{(0)} \\
&= \frac{B_x^{(0)}}{c}\frac{\partial \Phi}{\partial t}\left(t - \frac{x}{w_x^{(0)}}, y - \frac{w_y^{(0)}}{w_x^{(0)}} x\right) \\
&\quad + E_z^{(0)}\frac{\partial \Phi}{\partial y}\left(t - \frac{x}{w_x^{(0)}}, y - \frac{w_y^{(0)}}{w_x^{(0)}} x\right), \tag{A.28}
\end{aligned}$$

if the expressions (A.24)-(A.25) for B_n and $v_n - D$ are used. Here,

$$E_z^{(0)} = \frac{1}{c}\left(v_y^{(0)} B_x^{(0)} - v_x^{(0)} B_y^{(0)}\right). \tag{A.29}$$

All discontinuities cross the reconnection line at $x = 0$ and, therefore, the function $\Phi(t, y)$ is defined through the electric field along the reconnection line in its rest frame by evaluating (A.28) at $x = 0$,

$$\frac{B_x^{(0)}}{c}\frac{\partial \Phi}{\partial t}(t, y) + E_z^{(0)}\frac{\partial \Phi}{\partial y}(t, y) = E^*(t, y). \tag{A.30}$$

The solution of (A.30) is

$$\Phi(t, y) = \frac{c}{B_x^{(0)}} \int_0^t d\tau \, E^*\left[\tau, y - \frac{cE_z^{(0)}}{B_x^{(0)}}(t - \tau)\right]. \tag{A.31}$$

The function $f(t, x, y)$ describes the shape of the discontinuities and therefore it must be the same if (A.27) is evaluated at the different sides of the discontinuities. These conditions give the connections between the z-components of the displacement vector ζ in the different regions

$$\zeta_0 + \Phi_0(\mathbf{w}_A^{(0)}) = \zeta_1 + \Phi_1(\mathbf{w}_A^{(0)}), \quad (A), \tag{A.32}$$

$$\zeta_1 + \Phi_1(\mathbf{w}_S^{(0)}) = \zeta_2 + \Phi_2(\mathbf{w}_S^{(0)}), \quad (S), \tag{A.33}$$

$$\zeta_2 = \tilde{\zeta}_2, \quad (C), \tag{A.34}$$

$$\tilde{\zeta}_2 + \tilde{\Phi}_2(\tilde{\mathbf{w}}_S^{(0)}) = \tilde{\zeta}_1 + \tilde{\Phi}_1(\tilde{\mathbf{w}}_S^{(0)}), \quad (\tilde{S}), \tag{A.35}$$

$$\tilde{\zeta}_1 + \tilde{\Phi}_1(\tilde{\mathbf{w}}_A^{(0)}) = \tilde{\zeta}_0 + \tilde{\Phi}_0(\tilde{\mathbf{w}}_A^{(0)}), \quad (\tilde{A}), \tag{A.36}$$

with

$$\Phi_i(\mathbf{w}_j^{(0)}) \equiv \Phi_i\left(t - \frac{x}{w_{jx}^{(0)}}, y - \frac{w_{jy}^{(0)}}{w_{jx}^{(0)}}x\right), \tag{A.37}$$

evaluated with (A.31) in region i, and $j = A, S, \tilde{A}$, or $\tilde{S}$. Using this result, we are able to describe the inner structure of the boundary layer, which is determined by the reconnection electric field in combination with the initial current, corresponding to the change in the tangential magnetic field components.

12 References

Abe, S. A., and M. Hoshino (2001), Nonlinear evolution of plasmoid structure, *Earth Planets Space*, **53**, 663-671.

Akhiezer, A. I., I. A. Akhiezer, R. V. Polovin, A. G. Sitenko, and K. N. Stepanov (1975), *Plasma electrodynamics*, Vol. 1, Pergamon Press, Oxford.

Alfvén, H. (1943), On the existence of electromagnetic-hydrodynamic waves, *Ark. Mat. Astron. Fys.*, **29**, 1-7.

Alfvén, H. (1977), Electric currents in cosmic plasmas, *Rev. Geophys. Space Phys.*, **15**, 271-284.

Axford, W. I., and J. F. McKenzie (1996), Implications of observations of the solar wind and corona for solar wind models, *Astrophys. Space Sci.*, **243**, 1-4.

Balogh, A., et al. (2001), The Cluster magnetic field investigation: overview of in-flight performance and initial results, *Ann. Geophys.*, **19**, 1207-1217.

Barnes, A. (1983), Hydromagnetic waves, turbulence, and collisionless processes in the interplanetary medium, in *Solar–terrestrial physics*, eds. R. L. Carovillano and J. M. Forbes, D. Reidel, Dordrecht.

Biermann, L. (1951), Kometenschweife und solare Korpuskularstrahlung, *Z. Astrophys.*, **29**, 274-286.

Biernat, H. K., M. F. Heyn, and V. S. Semenov (1987), Unsteady Petschek reconnection, *J. Geophys. Res.* **92**, 3392-3396.

Biernat, H. K., V. S. Semenov, and R. P. Rijnbeek (1998), Time-dependent 3D Petschek-type reconnection: A case study for magnetopause conditions, *J. Geophys. Res.* **103**, 4693-4706.

Biernat, H. K., C. J. Farrugia, G. R. Lawrence, V. S. Semenov, D. F. Vogl, M. T. Kiendl, R. P. Rijnbeek, and N. V. Erkaev (2002), Flux transfer events at the magnetopause: Data aspects and theoretical approaches, *Recent Res. Devel. Plasmas* **2**, 1-17.

Blanchard, G. T., L. R. Lyons, O. de la Beaujardière, R. A. Doe, and M. Mendillo (1996), Measurments of the magnetotail reconnection rate, *J. Geophys. Res.*, *101*, 15265-15276.

Blanchard, G. T., L. R. Lyons, O. de la Beaujardière (1996), Magnetotail reconnection rate during magnetospheric substorms, *J. Geophys. Res.*, *102*, 24303-24312.

Borg, A. L., M. Øieroset, T. D. Phan, F. S. Mozer, A. Pedersen, C. Mouikis, J. P. McFadden, C. Twitty, A. Balogh, and H. Rème (2005), Cluster encounter of a magnetic reconnection diffusion region in the near–Earth magnetotail on September 19, 2003, *Geophys. Res. Lett.*, **32**, L19105.

Brio, M. and C. C. Wu. (1988), An upwind differencing scheme for the equations of ideal magnetohydrodynamics, *J. Comput. Phys.*, **75**, 400–422.

Cagniard, L. (1939), *Reflexion et Refraction des Ondes Seismiques Progressive*, Gauthier–Villard, Paris; English translation by Flinn, E. A., and C. H. Dix, (1962), *Reflection and refraction of progressive seismic waves*, McGraw–Hill, New York.

Cattell, C., J. Dombeck, J. Wygant, J. F. Drake, M. Swisdak, M. L. Goldstein, W. Keith, A. Fazakerley, M. Andre, E. Lucek, and A. Balogh (2005), Cluster observations of electron holes in association with magnetotail reconnection and comparison to simulations, *J. Geophys. Res.*, **110**, A01211, 10.1029/2004JA010519.

Chapman, S., and V. C. Ferraro (1930), A new theory of magnetic storms, *Nature*, **126**, 129–130.

Crooker, N. U. (1979), Dayside merging and cusp geometry, *J. Geophys. Res.*, **84**, 951-959.

Crooker, N. U., G. L. Siscoe, T. E. Eastman, L. A. Frank, and R. D. Zwickl (1984), Large–scale flow in the dayside magnetosheath, *J. Geophys. Res.*, **89**, 9711—9719.

Daly, P. W., and E. Keppler (1983), Remote sensing of a flux transfer event with energetic particles, *J. Geophys. Res.*, **88**, 3971–3980.

Daly, P. W., D. J. Williams, C. T. Russell, and E. Keppler (1981), Particle signatures of magnetic flux transfer events at the magnetopause, *J. Geophys. Res.*, **86**, 1628–1632.

Defouw, R. J. (1976), Wave propagation along a magnetic tube, *Astrophys. J.*, **209**, 266–269.

deHoop, A. T. (1960), The surface line source problem, *Appl. Sci. Res. B*, **8**, 349–356.

Dungey, J. W. (1953), Conditions for the occurrence of electrical discharges in astrophysical systems, *Phil. Mag.* **44**, 725–738.

Dungey, J. W. (1961), Interplanetary magnetic field and the auroral zones, *Phys. Rev. Lett.*, **6**, 47–48.

Eriksson, S., M. Øieroset, D. Baker, C. Mouikis, A. Vaivads, M. Dunlop, H. Rème, R. Ergun, and A. Balogh (2004), Walén and slow-mode shock analyses in the near-Earth magnetotail in connection with a substorm onset on 27 August 2001, *J. Geophys. Res.*, **109**, A05212, doi:10.1029/2003JA010534.

Erkaev, N. V. (1988), Results of the investigation of MHD flow around the magnetosphere (review), *Geomag. Aeron.*, **28**, 455–461.

Erkaev, N. V., V. S. Semenov, I. V. Alexeev, and H. K. Biernat (2001), Rate of steady–state reconnection in an incompressible plasma, *Phys. Plasmas* **8**, 4800–4809.

Farrugia, C. J., R. C. Elphic, D. J. Southwood, and S. W. H. Cowley (1987), Field and flow perturbations outside the reconnected field line region in flux transfer events: theory, *Planet. Space Sci.*, **35**, 227–240.

Farrugia, C. J., E. Lund, P. Sandholt, J. Wild, S. W. H. Cowley, A. Balogh, C. Mouikis, E. Moebius, M. W. Dunlop, J. Bosqued, C. Carlson, G. Parks, J. Cerisier, J. Kelly, J.–A. Sauvaud, and H. Réme (2004), Pulsed flows at the high–latitude cusp poleward boundary, and associated ionospheric convection and particle signatures, during a Cluster–FAST–SuperDARN–Søndrestrøm conjunction under a southwest IMF, *Ann. Geophys.*, **22**, 2891–2905.

Frey, H. U., T. D. Phan, S. A. Fuselier, and S. B. Mende (2003), Continuous magnetic reconnection at Earth's magnetopause, *Nature*, **426**, 533-537.

Fuselier, S. A., S. M. Petrinec, and K. J. Trattner (2000), Stability of the high-latitude reconnection site for steady northward IMF, *Geophys. Res. Lett.*, **27**, 473.

Fuselier, S. A., H. U. Frey, K. J. Trattner, S. B. Mende, and J. L. Burch (2002), Cusp aurora dependence on interplanetary magnetic field B_z, *J. Geophys. Res.*, **107**, A7, doi:10.1029/2001JA900165.

Fuselier, S. A., K. J. Trattner, S. M. Petrinec, C. J. Owen, and H. Rème (2005), Computing the reconnection rate at the Earth's magnetopause using two spacecraft observations, *J. Geophys. Res.*, **110**, A06212, doi:10.1029/2004JA010805.

Giovanelli, R. G. (1946), A theory of chromospheric flares, *Nature* **158**, 81–82.

Gonzalez, W. D. and F. S. Mozer (1974), A quantitative model for the potential resulting from reconnection with an arbitrary interplanetary magnetic field, *J. Geophys Res.*, **79**, 4186-4194.

Gosling, J. T., J. R. Asbridge, S. J. Bame, W. C. Feldman, G. Paschmann, N. Sckopke, and C. T. Russell (1982), Evidence for quasi–stationary reconnection at the dayside magnetopause, *J. Geophys. Res.*, **87**, 21472158.

Gosling, J. T., M. F. Thomsen, S. J. Bame, R. C. Elphic, and C. T. Russell (1991), Observations of reconnection of interplanetary and lobe magnetic field lines at the high-latitude magnetopause, *J. Geophys. Res.*, **96**, 14097-14106.

Grießmeier, J.–M., A. Stadelmann, T. Penz, H. Lammer, F. Selsis, I. Ribas, U. Motschmann, H. K. Biernat, and W. W. Weiss (2004), The effect of tidal locking on the stellar wind–magnetosphere interaction of

Hot Jupiters, *Astron. Astrophys.*, **425**, 753–762.

Hadamard, J. (1932), *Le problème de Cauchy et les équations aux dérivées partielles Lineaires Hyperboliques*, Hermann, Paris.

Haerendel, G., G. Paschmann, N. Sckopke, H. Rosenbauer, and P. C. Hedgecock (1978), The frontside boundary layer of the magnetopause and the problem of reconnection, *J. Geophys. Res*, *83*, 3195–3216.

Hanslmeier, A. (2002), *The sun and space weather*, Kluwer Academic Publ., Dordrecht.

Harten, A. (1997), High resolution schemes for hyperbolic conservation laws, *J. Comp. Phys.*, **135**, 260–278.

Harvey, C. C. (1998), Spatial gradients and the volumetric tensor, in *Analysis methods for multi–spacecraft data*, eds. G. Paschmann and P. W. Daly, ISSI SR–001, Bern, Switzerland, 307–322.

Hasegawa, H., B. U. Ö. Sonnerup, M. W. Dunlop, A. Balogh, S. E. Haaland, B. Klecker, G. Paschmann, B. Lavraud, I. Dandouras, and H. Rè me (2004), Reconstruction of two–dimensional magnetopause structures from Cluster observations: verification of method, *Ann. Geophys.*, **22**, 1251–1266.

Hasegawa, H., B. U. Ö. Sonnerup, B. Klecker, G. Paschmann, M. W. Dunlop, and H. Rè me (2005), Optimal reconstruction of magnetopause structures from Cluster data, *Ann. Geophys.*, **23**, 973–982.

Hau, L.–N., and B. U. Ö. Sonnerup (1999), Two–dimensional coherent structures in the magnetopause: Recovery of static equilibria from single-spacecraft data, *J. Geophys. Res.*, *104*, 6899–6918.

Heyn, M. F., and V. S. Semenov (1996), Rapid reconnection in compressible plasma, *Phys. Plasmas*, **3**, 2725–2741.

Heyn, M. F., H. K. Biernat, R. P. Rijnbeek, and V. S. Semenov (1988), The structure of reconnection layers, *J. Plasma Physics* **40**, 235–252.

Hones, E. W., Jr. (1977), Substorm processes in the magnetotail: Comments on "On hot tenuous plasma, fireballs, and boundary layers in the Earth's magnetotail" by L. A. Frank et al., *J. Geophys. Res.*, **82**, 5641–5643.

Hones, E. W., Jr. (1984). *Magnetic reconnection in space and laboratory plasmas*, Geophysical Monograph 30 (American Geophysical Union, Washington).

Hones, E. W., Jr., D. N. Baker, S. J. Bame, W. C. Feldman, J. T. Gosling, D. J. McComas, R. D. Zwickl, J. A. Slavin, E. J. Smith, and B. T. Tsurutani (1984), Structure of the magnetotail at 220 R_e and its response to geomagnetic activity, *Geophys. Res. Lett.*, **11**, 5-7.

Hu, Q., and B. U. Ö. Sonnerup (2001), Reconstruction of magnetic flux ropes in the solar wind, *Geophys. Res. Lett.*, *28*, 467–470.

Hu, Q., and B. U. Ö. Sonnerup (2003), Reconstruction of two-dimensional structures in the magnetopause: Method improvements, *J. Geophys. Res.*, **108**, 1011, doi:10.1029/2002JA009323.

Ieda, A., S. Machida, T. Mukai, Y. Saito, T. Yamamoto, A. Nishida, T. Terasawa, and S. Kokubun (1998), Statistical analysis of plasmoid evolution with GEOTAIL observations, *J. Geophys. Res.*, **103**, 4453–4466.

Kawano, H., and C. T. Russell (1997), Survey of flux transfer events observed with the ISEE 1 spacecraft: dependence on the interplanetary magnetic field, *J. Geophys. Res.* **102**, 11307–11314.

Kiendl, M. T., V. S. Semenov, I. V. Kubyshkin, H. K. Biernat, R. P. Rijnbeek, and B. P. Besser (1997), MHD analysis of Petschek–type reconnection in nonuniform field and flow geometries, *Space Sci. Rev.* **79**, 709–755.

Kippenhahn, R., and C. Möllenhoff (1975), *Elementare Plasmaphysik*, BI–Wissenschaftsverlag, Zürich.

Kistler, L., C. Mouikis, E. Moebius, B. Klecker, J.-A. Sauvaud, H. Rème, A. Korth, M. Marocci, R. Lundin, G. Parks, and A. Balogh (2005), Contribution of nonadiabatic ions to the cross-tail current in an O^+ dominated thin current sheet, *J. Geophys. Res.*, **110**, A06213, doi:10.1029/2004JA010653.

Kubyshkina, M. V., V. A. Sergeev, S. V. Dubyagin, S. Wing, P. T. Newell, W. Baumjohann, and A. T. Y. Liu (2002), Constructing the magnetospheric model including pressure measurements, *J. Geophys. Res.*, **107**, A6, SMP 4-1, doi:10.1029/2001JA900167.

Lamb, H. (1904), On the propagation of tremors over the surface of an elastic solid, *Philos. Trans. R. Soc. London Ser. A*, **203**, 1–42.

Lawrence, G. R., R. P. Rijnbeek, and V. S. Semenov (2000), Remote sensing of flux transfer events: Investigations of theoretical constraints based on model magnetopause time series data, *J. Geophys. Res.*, **105**, 7629–7638.

LeVeque, R. J. (2002), *Finite Volume Methods for Hyperbolic Problems*, Cambridge University Press, Cambridge.

Moldwin, M. B., and W. J. Hughes (1992), On the formation and evolution of plasmoids: A survey of ISEE 3 geotail data, *J. Geophys. Res.*, **97**, 19259–19282.

Mozer, F. S., S. D. Bale, T. D. Phan, and J. A. Osborne (2003), Observations of electron diffusion regions at the subsolar magnetopause, *Phys. Rev. Lett.*, **91**, id. 245002.

Nagai, T., I. Shinohara, M. Fujimoto, M. Hoshino, Y. Saito, S. Machida, and T. Mukai (2001), Geotail observations of the Hall current system:

Evidence for magnetic reconnection in the magnetotail, *J. Geophys. Res.*, *106*, 25929–25950.

Nakamura, R., W. Baumjohann, T. Nagai, M. Fujimoto, T. Mukai, B. Klecker, R. A. Treumann, A. Balogh, H. Rème, J.-A. Sauvaud, L. Kistler, C. Mouikis, C. J. Owen, A. Fazakerley, J. Dewhurst, and Y. Bogdanova (2004), Flow shear near the boundary of the plasma sheet observed by Cluster and Geotail, *J. Geophys. Res*, **109**, A05204, doi:10.1029/2003JA010174.

Nakamura, R., O. Amm, H. Laakso, N. Draper, M. Lester, A. Grocott, B. Klecker, I. McCrea, A. Balogh, H. Rème, and M. Andre (2005), Localized fast flow disturbance observed in the plasma sheet and in the ionosphere, *Ann. Geophys.*, **99**, 553–566.

Neff, J. E., T. W. Speiser, and D. J. Williams (1987), Magnetosheath quasi-trapped distributions and ion flows associated with reconnection, *J. Geophys. Res.*, **92**, 1177–1184.

Øieroset, M., P. E. Sandholt, W. F. Denig, and S. W. H. Cowley (1997), The northward IMF cusp aurora and high–latitude magnetopause reconnection, *J. Geophys. Res.*, **102**, 11349–11362.

Øieroset, M., T. D. Phan, M. Fujimoto, R. P. Lin, and R. P. Lepping (2001), In situ detection of collisionless reconnection in the Earth's magnetotail, *Nature*, **412**, 414–417.

Øieroset, M., R. P. Lin, T. D. Phan, D. E. Larson, and S. D. Bale (2002), Evidence for electron acceleration up to 300 keV in the magnetic reconnection diffusion region of Earths magnetotail, *Phys. Rev. Lett.*, **89**, id. 195001.

Orszag, S. A. and C.–M. Tang (1979), Small–scale structure of two-dimensional magnetohydrodynamic turbulence, *J. Fluid Mechanics*, **90**, 129–143.

Østgaard, N., J. Moen, S. B. Mende, H. U. Frey, T. J. Immel, P. Gallop, K. Oksavik, and M. Fujimoto (2005), Estimates of magnetotail reconnection rate based on IMAGE FUV and EISCAT measurements, *Ann. Geophys.*, *23*, 123–134.

Otto, A. (1991), Three–dimensional MHD computations of magnetic reconnection, in *Theoretical problems in space and fusion plasmas*, eds. H. K. Biernat, S. J. Bauer, and M. Heindler (Österreichische Akademie der Wissenschaften, Wien).

Owen, C. J., J. A. Slavin, A. N. Fazakerley, M. W. Dunlop, and A. Balogh (2005), Cluster electron observations of the separatrix layer during travelling compression regions, *Geophys. Res. Lett.*, **32**, L03104, 10.1029/2004GL021767.

Papamastorakis, I., G. Paschmann, W. Baumjohann, B. U. Ö. Sonnerup, and H. Lühr (1989), Orientation, motion, and other properties of flux transfer event structures on September 4, 1984, *J. Geophys. Res.*, **94**, 8852–8866.

Parker, E. N. (1957), Sweet's mechanism for merging magnetic fields in conducting fluids, *J. Geophys. Res.* **62**, 509–520.

Parker, E. N. (1958). Dynamics of the interplanetary gas and magnetic fields, *Astrophys. J.*, **128**, 664–676.

Paschmann, G., B. U. Ö. Sonnerup, I. Papamastorakis, N. Sckopke, G. Haerendel, S. J. Bame, J. R. Asbridge, J. T. Gosling, C. T. Russell, and R. C. Elphic (1979), Plasma acceleration at the Earth's magnetopause: evidence for reconnection, *Nature*, **282**, 243–246.

Penz, T. (2002), *Remote sensing of flux transfer events based on the theory of ill-posed inverse problems*, Diploma Thesis, University of Graz, Austria.

Penz, T., V. S. Semenov, V. V. Ivanova, I. B. Ivanov, V. A. Sergeev, R. Nakamura, M. F. Heyn, I. V. Kubyshkin, and H. K. Biernat (2004), Application of a reconstruction method for the reconnection rate applied to Cluster data from the Earth magnetotail, in *Proceedings of the 5th International Conference on Problems of Geocosmos*, St. Petersburg, Russia, 109–113.

Penz, T., V. V. Ivanova, V. S. Semenov, I. B. Ivanov, V. A. Sergeev, R. Nakamura, H. K Biernat, I. V. Kubyshkin, and M. F. Heyn (2005), Reconstruction of nightside flux transfer events using Cluster data, in *Proceedings of the Workshop on Auroral Phenomena*, Apatity, Russia, 44–47.

Penz, T., V. S. Semenov, V. V. Ivanova, H. K. Biernat, V. A. Sergeev, R. Nakamura, I. V. Kubyshkin, I. B. Ivanov, and M. F. Heyn (2006a), A reconstruction method for the reconnection rate applied to Cluster magnetotail measurements, *Adv. Space Res.*, in press.

Penz, T., V. V. Ivanova, V. S. Semenov, R. Nakamura, M. F. Heyn, I. B. Ivanov, I. V. Kubyshkin, and H. K. Biernat (2006b), Magnetic reconnection in the Earth's magnetotail: Reconstruction method and data analysis, in *Space Science: New Research*, Nova Science Publisher, Hauppauge, NY, USA, in press.

Penz, T., V. S. Semenov, V. V. Ivanova, M. F. Heyn, H. K. Biernat, and I. B. Ivanov (2006c), Green's function of compressible Petschek–type magnetic reconnection, *Phys. Plasmas*, accepted.

Perthame, B. and M. Westdickenberg (2005), Total oscillation diminishing property for scalar conservation laws, *Numerische Mathematik*, **100**, 331–349.

Petschek, H. E. (1964), Magnetic field annihilation, in *Physics of solar flares*, ed. W. N. Hess, NASA Spec. Publ. 50, 425–440.

Phan, T. D., L. M. Kistler, B. Klecker, G. Haerendel, G. Paschmann, B. U. Ö. Sonnerup, W. Baumjohann, M. B. Bavassano–Cattaneok, C. W. Carlson, A. M. DiLellisk, K.–H. Fornacon, L. A. Frank, M. Fujimoto, E. Georgescu, S. Kokubun, E. Moebius, T. Mukai, M. Øieroset, W. R. Paterson, and H. Remé (2000), Extended magnetic reconnection at the Earth's magnetopause from detection of bi–directional jets, *Nature*, **404**, 848–850.

Phan, T. D., H. U. Frey, S. Frey, L. Peticolas, S. Fuselier, C. Carlson, H. Rème, J.–M. Bosqued, A. Balogh, M. Dunlop, L. Kistler, C. Mouikis, I. Dandouras, J.–A. Sauvaud, S. Mende, J. Mc–Fadden, G. Parks, E. Moebius, B. Klecker, G. Paschmann, M. Fujimoto, S. Petrinec, M. F. Marcucci, A. Korth, and R. Lundin (2003), Simultaneous Cluster and IMAGE observations of cusp reconnection and auroral proton spot for northward IMF, *Geophys. Res. Lett.*, **30**, doi:10.1029/2003GL016885.

Pinnock, M., G. Chisham, I. J. Coleman, M. P. Freeman, M. Hairston, and J.–P. Villain (2003), The location and rate of dayside reconnection during an interval of southward interplanetary magnetic field, *Ann. Geophys.*, **21**, 1467–1482.

Priest, E. R. (1984), Magnetic reconnection at the Sun, in *Magnetic reconnection in space and laboratory plasmas*, ed. E. W. Hones, Jr. (American Geophysical Union, Washington).

Priest, E. R., and T. G. Forbes (1986), New models for fast steady state magnetic reconnection, *J. Geophys. Res.* **91**, 5579–5588.

Priest, E. R., and T. G. Forbes (2000), *Magnetic reconnection: MHD theory and applications*, Cambridge University Press, Cambridge.

Pudovkin, M. I., and V. S. Semenov (1977), Peculiarities of the MHD flow by the magnetopause and generation of the electric field in the magnetosphere, *Ann. Geophys.*, **33**, 423–427.

Rème, H., et al. (2001), First multispacecraft ion measurements in and near the Earth's magnetosphere with the identical Cluster ion spectrometry (CIS) experiment, *Ann. Geophys.*, **19**, 1303–1354.

Retinò, A., M. B. Bavassano Cattaneo, M. F. Marcucci, A. Vaivads, M. André, Y. Khotyaintsev, T. Phan, G. Pallocchia, H. Rème, E. Moebius, B. Klecker, C.W. Carlson, M. McCarthy, A. Korth, R. Lundin, and A. Balogh (2005), Cluster multispacecraft observations at the high–latitude duskside magnetopause: implications for continuous and component magnetic reconnection, *Ann. Geophys.*, **23**, 461–473.

Rijnbeek, R. P. (1984), Flux transfer events: impulsive reconnection signatures at the Earth's magnetopause, Ph. D. thesis, University of London, England.

Rijnbeek, R. P., S. W. H. Cowley, D. J. Southwood, and C. T. Russell (1982), Observations of reverse polarity flux transfer events at the Earth's dayside magnetopause, *Nature* **300**, 23–26.

Runov, A., R. Nakamura, W. Baumjohann, R. A. Treumann, T. L. Zhang, M. Volwerk, Z. Vörös, A. Balogh, K.–H. Glassmeier, B. Klecker, H. Rème, and L. Kistler (2003), Current sheet structure near magnetic X–line observed by Cluster, *Geophys. Res. Lett.*, **30**, 1579, doi:10.1029/2002GL016730.

Russell, C. T., and R. C. Elphic (1978), Initial ISEE magnetometer results: magnetopause observations, *Space Sci. Rev.*, **22**, 681–715.

Saunders, M. (1991), The Earth's magnetosphere, in *Advances in Solar System Magnetohydrodynamics*, eds. E. R. Priest and A. W. Hood, Cambridge University Press, 357–398.

Schindler, K. (1974), A theory of the substorm mechanism, *J. Geophys. Res.*, **79**, 2803–2810.

Scholer, M. (1989), Undriven magnetic reconnection in an isolated current sheet, *J. Geophys. Res.* **94**, 8805–8812.

Schwartz, S. J. (1985), Solar wind and the Earth's bow shock, in *Solar system magnetic fields*, ed. E. R. Priest, D. Reidel Publishing Company, Dordrecht.

Schwenn, R. (1991), Der Sonnenwind, in *Plasmaphysik im Sonnensystem*, eds. K.–H. Glassmeier, and M. Scholer, BI–Wissenschaftsverlag, Mannheim.

Sergeev, V. A., V. S. Semenov, and M. V. Sidneva (1987), Impulsive reconnection in the magnetotail during substorm expansion, *Planet. Space Sci.*, **35**, 1199–1212.

Sergeev, V. A., R. C. Elphic, F. S. Mozer, A. Saint–Marc, and J.–A. Sauvaud (1992), A two–satellite study of nightside flux transfer events in the plasma sheet, *Planet. Space Sci.*, **40**, 1551–1572.

Sergeev, V. A., M. V. Kubyshkina, W. Baumjohann, R. Nakamura, O. Amm, T. Pulkkinen, V. Angelopoulos, S. B. Mende, B. Klecker, T. Nagai, J.–A. Sauvaud, J. A. Slavin, and M. F. Thomsen (2005), Transition from substorm growth to substorm expansion phase as observed with a radial configuration of ISTP and Cluster spacecraft, *Ann. Geophys.*, **23**, 2183-2198.

Semenov, V. S., M. F. Heyn, and I. V. Kubyshkin (1983), Reconnection of magnetic field lines in a nonstationary case, *Sov. Astron.* **27**, 660–665.

Semenov, V. S., E. P. Vasilyev, and A. I. Pudovkin (1984), A scheme for the nonsteady reconnection of magnetic lines of force, *Geomagn. Aeron.* **24**, 448–452.

Semenov, V. S., I. V. Kubyshkin, V. V. Lebedeva, R. P. Rijnbeek, M.

F. Heyn, H. K. Biernat, and C. J. Farrugia (1992), A comparison and review of steady–state and time–varying reconnection, *Planet. Space Sci.*, **40**, 63–87.

Semenov, V. S., V. V. Lebedeva, H. K. Biernat, M. F. Heyn, R. P. Rijnbeek, and C. J. Farrugia (1995), Time–varying reconnection: Implications for magnetopause observations, *J. Geophys. Res.* **100**, 21779–21790.

Semenov, V. S., O. A. Drobysh, and M. F. Heyn (1998), Analysis of time–dependent reconnection in compressible plasmas, *J. Geophys. Res.*, **103**, 11863–11873.

Semenov, V. S., M. F. Heyn, and I. B. Ivanov (2004a), Magnetic reconection with space and time varying reconnection rates in a compressible plasma, *Phys. Plasmas*, **11**, 62–70.

Semenov, V. S., I. V. Kubyshkin, R. P. Rijnbeek, and H. K. Biernat (2004b), Analytical theory of unsteady Petschek–type reconnection, in *Physics of Magnetic Reconnection in High–Temperature Plasmas*, ed. M. Ugai, Research Signpost, Trivandrum, India, 35–68.

Semenov, V. S., T. Penz, V. V. Ivanova, V. A. Sergeev, H. K. Biernat, R. Nakamura, M. F. Heyn, I. V. Kubyshkin, and I. B. Ivanov (2005a), Reconstruction of the reconnection rate from Cluster measurements: First results, *J. Geophys. Res.*, **110**, A11217, doi:10.1029/2005JA011181.

Semenov, V. S., T. Penz, M. F. Heyn, I. B. Ivanov, I. V. Kubyshkin, H. K. Biernat, V. V. Ivanova, and R. P. Rijnbeek (2005b), Reconstruction of the reconnection rate from magnetic field disturbances in an incompressible plasma, in *Solar–planetary relations*, eds. by H. K. Biernat, H. Lammer, D. F. Vogl, and S. Mühlbachler, Research Signpost, Trivandrum, India, 261–302.

Shue, J.-H., J. K. Chao, H. C. Fu, C. T. Russell, P. Song, K. K. Khurana, and H. J. Singer (1997), A new functional form to study the solar wind control of the magnetopause size and shape, *J. Geophys. Res.*, **102**, 9497–9512.

Siscoe, G. L., N. F. Ness, and C. M. Yeates (1975). Substorms on Mercury, *J. Geophys. Res.* **80**, 4359–4363.

Slavin, J. A., R. P. Lepping, J. Gjerloev, D. H. Fairfield, M. Hesse, C. J. Owen, M. B. Moldwin, T. Nagai, A. Ieda, and T. Mukai (2003), Geotail observations of magnetic flux ropes in the plasma sheet, *J. Geophys. Res.*, **108**, doi:10.1029/2002JA009557.

Sonnerup, B. U. Ö, G. Paschmann, I. Papamastorakis, N. Sckopke, G. Haerendel, S. J. Bame, J. R. Asbridge, J. T. Gosling, and C. T. Russell (1981), Evidence for magnetic field reconnection at the Earth's magnetopause, *J. Geophys. Res.*, **86**, 10049–10067.

Sonnerup, B. U. Ö., H. Hasegawa, and G. Paschmann (2004), Anatomy of a flux transfer event seen by Cluster, *Geophys. Res. Lett.*, **31**, L11803, doi:10.1029/2004GL020134.

Southwood, D. J. (1985), Theoretical aspects of ionosphere–magnetosphere–solar wind coupling, *Adv. Space Res.*, **5**, 7–14.

Southwood, D. J., C. J. Farrugia, and M. A. Saunders (1988), What are flux transfer events?, *Planet. Space Sci.* **36**, 503–508.

Spreiter, J. R., A. L. Summers, and A. Y. Alksne (1966), Plasma flow around the magnetosphere, *Planet. Space Sci.*, **14**, 223–253.

Sweet, P. A. (1958), The neutral point theory of solar flares, in *Electromagnetic phenomena in cosmical physics*, IAU Symp. 6, ed. B. Lehnert, Cambridge Univ. Press, London.

Tikhonov, A. N., and V. Y. Arsenin (1977), *Solutions of ill-posed problems*, John Wiley, New York.

Tikhonov, A. N., and A. V. Goncharsky (1987), *Ill–posed problems in the natural sciences*, Mir Publishers, Moscow.

Toth, G. (1996), Comparison of some flux corrected transport and total variation diminishing numerical schemes for hydrodynamic and magnetohydrodynamic problems, *J. Comp. Phys.*, **128**, 82–100.

Toth, G. (2000), The $\nabla \cdot \mathbf{B} = 0$ constraint in shock–capturing magnetohydrodynamics codes, *J. Com. Phys.*, **161**, 605–652.

Trattner, K. J., S. A. Fuselier, and S. M. Petrinec (2004), Location of the reconnection line for northward interplanetary magnetic field, *J. Geophys. Res.*, **109**, 3219-3229.

Twitty, C., T. D. Phan, G. Paschmann, B. Lavraud, H. Rème, and M. Dunlop (2004), Cluster survey of cusp reconnection and its IMF dependence, *Geophys. Res. Lett.*, **31**, L19808, doi:10.1029/2004GL020646.

Ugai, M. (1992), Computer studies on development of the fast reconnection mechanism for different resistivity models, *Phys. Fluids. B. Plasma Phys.*, **4**, 2953–2963.

Ugai, M. and T. Tsuda (1977), Magnetic field–line reconnection by localized enhancement of resistivity. Part 1. Evolution in a compressible MHD fluid, *J. Plasma Phys.*, **17**, 337-356.

Vaivads, A., Y. Khotyaintsev, M. André, A. Retinò, S. C. Buchert, B. N. Rogers, P. Décréau, G. Paschmann, and T. D. Phan (2004), Structure of the magnetic reconnection diffusion region from four–spacecraft observations, *Phys. Rev. Lett.*, **93**, id. 105001.

Vasyliunas, V. M. (1975), Theoretical models of magnetic field line merging, 1, *Rev. Geophys. Space Sci.* **13**, 303–336.

Walthour, D. W., B. U. Ö. Sonnerup, G. Paschmann, H. Lühr, D. Klumpar, and T. Potemra (1993), Remote sensing of two–dimensional magnetopause structures, *J. Geophys. Res.*, **98**, 1489–1504.

Walthour, D. W., B. U. Ö. Sonnerup, R. C. Elphic, and C. T. Russell (1994), Double vision: Remote sensing of a flux transfer event with ISEE 1 and 2, *J. Geophys. Res.*, **99**, 8555–8563.

Wild, J. A., S. W. H. Cowley, J. A. Davies, H. Khan, M. Lester, S. E. Milan, G. Provan, T. K. Yeoman, A. Balogh, M. W. Dunlop, K.-H. Fornacon, and E. Georgescu (2001), First simultaneous observations of flux transfer events at the high–latitude magnetopause by the Cluster spacecraft and pulsed radar signatures in the conjugate ionosphere by the CUTLASS and EISCAT radars, *Ann. Geophys.*, **19**, 1491–1508.

Wild, J. A., S. E. Milan, S. W. H. Cowley, J. M. Bosqued, H. Rè me, T. Nagai, S. Kokubun, Y. Saito, T. Mukai, J. A. Davies, B. M. A. Cooling, A. Balogh, and P. W. Daly (2005), Simultaneous in–situ observations of the signatures of dayside reconnection at the high– and low–latitude magnetopause, *Ann. Geophys.*, **23**, 445–460.

Wygant, J. R., C. A. Cattell, R. Lysak, Y. Song, J. Dombeck, J. McFadden, F. S. Mozer, C. Carlson, G. Parks, E. A. Lucek, A. Balogh, M. Andre, H. Rème, M. Hesse, and C. Mouikis (2005), Cluster observations of an intense normal component of the electric field at a thin reconnecting current sheet in the tail and its role in the shock–like acceleration of the ion fluid into the separatrix region, *J. Geophys. Res.*, **110**, A09206, doi:10.1029/2004JA010708.

Yamada, M., H. Ji, S. Hsu, T. Carter, R. Kulsrud, N. Bretz, F. Jones, Y. Ono, and F. Perkins (1997), Study of driven magnetic reconnection in a labratory plasma, *Phys. Plasmas* **4**, 1936–1944.

Zong, Q.-G., T. A. Fritz, Z. Y. Pu, S. Y. Fu, D. N. Baker, H. Zhang, A. T. Lui, I. Vogiatzis, K.-H. Glassmeier, A. Korth, P. W. Daly, A. Balogh, and H. Rème (2004), Cluster observations of earthward flowing plasmoid in the tail, *Geophys. Res. Lett.*, **31**, L18803, doi:10.1029/2004GL020692.

Südwestdeutscher Verlag für Hochschulschriften

Wissenschaftlicher Buchverlag bietet
kostenfreie
Publikation
von
Dissertationen und Habilitationen

Sie verfügen über eine wissenschaftliche Abschlußarbeit zu aktuellen oder zeitlosen Fragestellungen, die hohen inhaltlichen und formalen Anspruchen genügt, und haben **Interesse an einer honorarvergüteten Publikation?**

Dann senden Sie bitte erste Informationen über Ihre Arbeit per Email an:
info@svh-verlag.de.

Unser Außenlektorat meldet sich umgehend bei Ihnen.

Südwestdeutscher Verlag für Hochschulschriften
Aktiengesellschaft & Co. KG
Dudweiler Landstr. 99
D – 66123 Saarbrücken
www.svh-verlag.de

Printed by Books on Demand GmbH, Norderstedt / Germany